Management and Industrial Engineering

Series Editor

J. Paulo Davim, Department of Mechanical Engineering, University of Aveiro, Aveiro, Portugal

This series fosters information exchange and discussion on management and industrial engineering and related aspects, namely global management, organizational development and change, strategic management, lean production, performance management, production management, quality engineering, maintenance management, productivity improvement, materials management, human resource management, workforce behavior, innovation and change, technological and organizational flexibility, self-directed work teams, knowledge management, organizational learning, learning organizations, entrepreneurship, sustainable management, etc. The series provides discussion and the exchange of information on principles, strategies, models, techniques, methodologies and applications of management and industrial engineering in the field of the different types of organizational activities. It aims to communicate the latest developments and thinking in what concerns the latest research activity relating to new organizational challenges and changes world-wide. Contributions to this book series are welcome on all subjects related with management and industrial engineering. To submit a proposal or request further information, please contact Professor J. Paulo Davim, Book Series Editor, pdavim@ua.pt

Petr Baron · Marek Kočiško · Anton Panda

Application of Troubleshooting Tools in the Monitored Production Processes

Springer

Petr Baron
Department of Computer Aided
Manufacturing Technologies
Technical University of Košice
Prešov, Slovakia

Marek Kočiško
Department of Computer Aided
Manufacturing Technologies
Technical University of Košice
Prešov, Slovakia

Anton Panda
Department of Automobile
and Manufacturing Technologies
Technical university of Košice
Prešov, Slovakia

ISSN 2365-0532 ISSN 2365-0540 (electronic)
Management and Industrial Engineering
ISBN 978-3-031-41427-5 ISBN 978-3-031-41428-2 (eBook)
https://doi.org/10.1007/978-3-031-41428-2

This Springer imprint is published by the registered company Springer Nature Switzerland AG
The registered company address is: Gewerbestrasse 11, 6330 Cham, Switzerland

Preface

The monograph is devoted to practical examples of troubleshooting tools applied in real technological processes. First and foremost, the monograph is intended for employees dealing with the issue of setting up technical systems and reliable operation of rotary machines.

The operational reliability of machines and equipment largely depends on the bearing condition and its service life. Rolling bearings are important parts of machines, mechanisms, and devices that enable them to operate reliably and provide for the safe and long-term functioning of the technical systems without friction and enormous wear. On the other hand, roller bearing damage is an important and significant phenomenon that can lead to the failure of machines, mechanisms, and equipment in all industrial areas where they are used. The analyses of the breakdown causes, their possible consequences, and their removal create conditions for ensuring the quality and reliability of rolling bearings during their technical life. Similar to other machine components, even in the case of roller bearings, there may be a variety of causes of premature damage and failure of the bearing. It is necessary to differentiate between bearing durability, determined by wear due to the application of load at operating speed, and bearing lifetime, which is a period of the bearing serviceability before it is for various reasons decommissioned. The life of the bearing is affected, for example, by improper installation, misalignment of bearings, manufacturing defaults in the production of connecting parts, improper handling of bearings, impurities deposited in bearings, or poor lubrication. When damage or other deficiencies occur in bearings, it is important to determine the cause so that measures can be taken to prevent further damage.

If we disregard the proper bearing handling, its installation, and the ideal lubrication regime, it is the dynamic forces in operation that have a significant impact on bearing life; they can also be caused by other components, their failures, and operational and technological impacts. In this context, the increasingly demanding operating conditions need to be taken into account. Not only do they cause that new materials applied in tribological nodes are developed but also that technical diagnostic tools are applied so that bearing defects that are often hidden can be early identified.

The aim of bearing vibrodiagnostics is not only to monitor the mere bearing condition but also to identify hidden component defects that have a significant effect on bearing wear.

The monograph was supported by grants VEGA 1/0026/22, VEGA 1/0226/21, and KEGA 002TUKE-4/2023.

Vote of Thanks

The author expresses gratitude to reviewers for valuable monograph and substantive and formal observations that raise the overall level of quality publications.

Publisher/Editor: Springer International Publishing, Switzerland

Edition Scientific and Technical Literature: Monograph

Reviewers: Dr.h.c. prof. dr.hab.Inž. Tadeusz Zaborowski
prof. Ing. Jozef Pilc, CSc.
prof. dr hab. inż. Stanisław Legutko
Ing. Jozef Mikita, PhD.

Prešov, Slovakia

Petr Baron
Marek Kočiško
Anton Panda

Contents

About the Authors

Petr Baron doc. Ing. Ph.D.: University study—Final state examination, title Engineer Master of Science. Faculty of Mechanical Engineering, TU Kosice, Slovakia (1994); doctoral study—title Ph.D. Technical University in Kosice, Faculty of Manufacturing technology with the seat in Presov, Slovakia (1998); 2007—to present: Associate Professor, Faculty of Manufacturing Technologies with the seat in Presov, Slovakia. Professional profiling is focused on computer support of engineering activities in pre-production stages, computer simulation of production systems and processes, simultaneous engineering and risk management of technical systems. As part of the educational process, he is a guarantor of subjects in the 1st and 2nd level of university studies for the Computer Support of Production Technologies study program and also a trainer of Ph.D. students both in full and part-time form. He has good knowledge of IT and implementation of IT tools in technology workplaces. He is the author of several utility models and a number of scientific publications published in journals processed by ISI Current Contents, available through the Web of Knowledge portal. AWARDS (CERTIFICATES): Awarded as project leader of the project VEGA MŠ SR č. VEGA 1/0013/11—Innovation of methodology of risk identification and valuation process of undesirable events on technological workplaces, Project duration: 2011–2015, project leader—Awarded as project leader of the project KEGA MŠ SR č. 3/7167/09—a certificate based on the results of final evaluation of KEGA project of the Ministry of Education of the Slovak Republic, Proposal of Interactive Educational Manual for Systems Areas of the Computer Aided Technological Preparation of Production. Project duration: 2009–2011, project leader—Awarded as project leader of the project KEGA MŠ SR č. KEGA 3/4135/06—a certificate based on the results of final evaluation of KEGA project of the Ministry of Education of the Slovak Republic, Creating of electronic educational manuals for risk management in manufacturing technologies. Project duration 2006–2008, project leader—Awarded as project leader of the project VEGA MŠ SR č. VEGA 1/6199/99—a certificate based on the results of final evaluation of VEGA project of the Ministry of Education of the Slovak Republic, Bionic production systems, their hierarchy, arrangement and application in a modern enterprise. Project duration: 1999–2001.

Marek Kočiško Prof. Ing., Ph.D.: graduated in 2001 from the Faculty of Manufacturing Technologies of the Slovak Technical University of Kosice and he completed his Ph.D. (2004) in the field of study Engineering technologies and materials. He currently works as a professor position (2022) and the head of the Department of Computer Aided of Manufacturing Technologies at the Faculty of Manufacturing Technologies with a seat in Presov, TUKE. His work focuses on computer aided of production processes, CAD/CAM/CAE systems, simulation of production systems, computer networks, virtual (augmented) reality as well as reverse and simultaneous engineering. His strengths include knowledge of PC hardware, computer networks, notification technology, excellent knowledge of operating systems, office systems, computer simulation systems and CAD/CAM/CAE systems (Creo, Unigraphics NX, Solid Edge, etc.). He is an expert member of many national and international associations and also a reviewer member of many journals registered in Current Content, Web of Science and Scopus databases.

Anton Panda Prof. Ing., Ph.D.: University studies—Faculty of Mechanical Engineering TU Košice (Ing.-1987); terminated doctoral studies—Faculty of manufacturing technologies TU Košice (Ph.D.-2002), associate professor of study branch 5.2.51 manufacturing technologies, FMT TU Košice (assoc. prof.-2008), professor of study branch 5.2.51 manufacturing technologies, FMT TU Košice (prof.-2015). 29 years of experience in the engineering company supplying the products for demanding automotive, also farm and agricultural industry (constructor of special purpose machinery and equipment, systems analyst, head the department of development and technical preparation of production, methodist of statistical methods, commercial and technical director, director of quality). In the present expertise and design activities in the area of development, production and verification of rolling bearings, in the areaof deposition with rolling bearings for various domestic and foreign customers. Since 2008 (since 1994 external) operates as pedagogue and scientist at the Faculty of manufacturing technologies TU Košice with the seat in Prešov, as well as an expert—coordinator (auditor) of quality management systems. He is the author (co-author) of 17 monographs (11 foreign, 6 domestic)—of it 3 monograph in Springer publishing, 2 university textbooks (1 foreign, 1 domestic), 16 university lecture notes, author's certificates (16), patents and discoveries (15), catalogs of bearings (2), several domestic and foreign original scientific papers in the scientific and professional journals, in Current Contents Connect journals in Web of Science (21), in impacted journals and publications led in the world renowned databases (Web of Science—117, Scopus—148) and in proceedings from domestic and foreign scientific conferences from the following areas: Automobile production, manufacturing technologies, experimental methods in the manufacturing technologies, machining, development, manufacturing and verification of new products in accordance with the standards EN ISO 9001 and in accordance to the specific requirements of automobile manufacturers IATF 16 949, quality control, statistical methods and techniques of quality for the production of parts, capability of machine, capability of manufacturing processes, capability of gauges and measuring equipment, technical preparation of production, product audits, system audits of quality management system, analysis of potential errors and their effects on construction (FMEA-K) and on manufacturing process/technology (FMEA-V), statistical regulation of manufacturing processes SPC, process of approval of parts to the production PPAP,modern quality planning of product APQP, control plans and regulation, requirements the association of automobile manufacturers in Germany VDA 6.1, quality system requirements for suppliers of Ford, Chrysler, GM, specific requirements the using of EN ISO 9001:2015 inorganizations ensuring the mass production in automotive industry IATF 16949, method of Poka-Yoke, quality assurance before the mass production for suppliers of automobile manufacturers in Germany VDA 4.3, quality assurance of supplies for suppliers of automobile manufacturers in Germany VDA 2, product liability, method of Global 8D (8-step method for solving of problems), etc. At these works are registered the various domestic and foreign quotations and testimonials in the worldwide databases. Solver of several projects and grant projects for engineering companies at home and abroad, solver of research tasks, author of the directives, methodological guidelines, technical regulations and other technical documentation for domestic and foreign manufacturing companies. He is auditor of quality system management on Technical University in Košice, Slovakia. Active collaboration with the university workplaces at home and abroad. He is recognized as an expert for the production of bearings in companies in Germany, Italy, China, Slovakia and Czech Republic. As the coordinator of research collective and co-author of documentation EFQM has won the Award for improvement of performance in the competition National award of Slovak Republic for quality in the year 2010 for the Technical University of Košice. In the same competition, he has won the same award in year 2012, when the Technical University of Košice has obtained the highest score in its category. Since 2014, he is a member of the Polish Academy of Sciences. Since 2014, he is a member of the ASME, USA.

Symbols and Abbreviations

ACC	ACCeleration
AV	Average Value
BPFI	Ball Pass Frequency Inner
BPFO	Ball Pass Frequency Outer
BSF	Ball Spin Frequency
CCT IR	Total amount of Dirt Particles in the Process Medium—Fe
CLA	Center-Line Average
ČSN	Czech Technical Standards
dB	Unit Decibel
D_L	Number of Large Particles
D_S	Number of Small Particles
En	Envelope
EnvAcc	Envelope Acceleration
FFT	Fast Fourier Transform
gE	Unit of Envelope Acceleration
HF	High-Frequency Emission
HFD	High-Frequency Detection
ISO	International Organization for Standardization
LF	Low-Frequency Emission
LM	Lost Motion
PAO	Polyalphaolefin
PG	Propylene Glycol
PLP	Percent of Large Particles
PPM	Parts Per Million
PtP	Peak to Peak
PV	Peak Value
Ra	Average Roughness
RMS	Root Mean Square
Rp	Maximum Profile Peak Height
RPM	Rev. Per Min
Rv	Maximum Profile Valley Depth

Rz	Ten Point Height of Irregularities
SEE	Spectral Emitted Energy
S (t)	State of the System at Time t
SKF	Acoustic Emission Enveloping (AEE)
STN	Slovak Technical Standards
Vel	Velocity
XRF	X-Ray Fluorescence
WPC	Wear Particles Concentration

Chapter 1
Introduction—Proactive Maintenance Methods

Proactive maintenance methods represent an important tool not only of cost management, but also of protecting all elements in the system of human–machine environment. The basic idea of modern maintenance is to move maintenance from the "ex post" event stage to the "ex ante" stage. Thus, maintenance planning and methods of preventive, predictive, and proactive maintenance are applied. When maintaining technical systems, it is important to ensure trouble-free operation as long as possible and to maximize the life of machinery and equipment. In preventive maintenance, this goal is achieved by setting regular service intervals, during which minor repairs and inspections of the relevant equipment are performed. However, a problem with such approach is to determine an optimal service interval. Frequent service interventions increase maintenance costs and also have an impact on production (should they require shutting down of the production equipment) [1, 2].

Predictive maintenance seeks to identify the right moment for maintenance using advanced statistical methods and, nowadays, using the tools of artificial intelligence. Unlike preventive maintenance, the maintenance of each device is assessed and planned based on the current state of the device, with various models estimating the time and date to failure. Subsequently, it is possible to extend or shorten maintenance cycles according to the real state in which the equipment is found [1, 2]. Thus, the predictive maintenance approach is to repair only those things on machines and equipment that are absolutely necessary to be repaired at the time when such repair is necessary.

Predictive maintenance allows [1, 3, 4]:

- to predict failure
- to identify unusual system behavior
- to identify incorrectly set operating parameters
- to estimate the remaining life of the technical system

P. Baron et al., *Application of Troubleshooting Tools in the Monitored Production Processes*, Management and Industrial Engineering,
https://doi.org/10.1007/978-3-031-41428-2_1

Predictive maintenance requires the so-called Condition Monitoring, i.e., monitoring symptoms of developing defects. Such approach requires technical troubleshooting tools be applied. Its effectiveness depends on the people, the technologies used, and also the instrumentation [1, 5]. Methods of technical troubleshooting, such as thermodiagnostics, tribodiagnostics, vibration diagnostics, defectoscopy, etc., can be included here.

Proactive maintenance eliminates the causes of machine damage. It is this reason that makes the use of the means of technical troubleshooting indispensable. Vibration troubleshooting plays an important role in determining the current state of rotary machines.

Modern maintenance strategy prefers proactive approach. Proactive maintenance is based on continuous activity, involving the monitoring and management of the basic causes of the malfunctions. It represents a strategy that uses corrective action to prevent the occurrence of malfunctions, whereas these activities are focused on the causes of malfunctions. This includes preventive as well as predictive maintenance. This can be used to achieve an environment that supports extended service life of the technical system. Proactive maintenance method is used to eliminate unnecessary maintenance, unnecessary general overhauls, unscheduled down time and excessive stock of spare parts. Across all industrial segments, proactive maintenance methods presently save thousands or even millions of dollars on machinery maintenance costs every year. This concept of saving large volumes of maintenance may often be considered erroneous. According to DuPont, "maintenance represents the biggest individual cost in their plants: In many companies, this cost often exceeds the net annual profit." Considering that up to 90% of maintenance work in some companies involves costly (and adversely affecting productivity) repairs after malfunctions, it is easy to see the benefits of such maintenance strategy. Maintenance of consequent (post-malfunction) and preventive type involves significant financial requirements. Therefore, predictive maintenance is being implemented. This concept holds that repairs of machinery should be performed only on those parts and only when absolutely necessary. Predictive maintenance requires condition monitoring, i.e., monitoring of the symptoms of developing faults. This type of maintenance uses instruments of technical diagnostics, implemented across all pieces of machinery within the plant. Its effectiveness is greatly dependent on human factor, used technology and used instrumentation. The processes include vibration diagnostics, tribology, thermo-diagnostics, defectoscopy, etc. A successful implementation may bring cost savings of up to one half of standard maintenance costs. Long-term analysis discovered that most failures repeat and have unique underlying cause [5–8].

Proactive maintenance involves all components of predictive maintenance, focusing not only on the current symptoms of machine condition (e.g., failed bearings), but also on the search and elimination of the causes of this undesirable condition (e.g., bearing failed due to improper set-up of the machine).

References

1. Baron, P. - Kočiško, M. - Hlavatá, S. - Franas, E.: Vibrodiagnostics as a predictive maintenance tool in the operation of turbo generators of a small hydropower - 2022. In: Advances in Mechanical Engineering. - Thousand Oaks (USA) : SAGE Publications vol. 14, no. 5 (2022), pp. 1–15 [online, print]. https://doi.org/10.1177/16878132221101023.
2. Panda, A. - Prislupčák, M.: Valuation of vibrations during machining, Trends and Innovative Approaches in Business Processes, 17. International Scientific Conference, Almanac, Košice, 2014, pp. 1–5.
3. Krolczyk, G. - Gajek, M. and Legutko, S.: Predicting the tool life in the dry machining of duplex stainless steel, Eksploatacja i Niezawodnosc-Maintenance and Reliability, vol. 15, no. 1, pp. 62–65, 2013.
4. Krolczyk, G. - Krolczyk, J. - Legutko, S. and Hunjet, A.: Effect of the disc processing technology on the vibration level of the chipper during operations process, Technical Gazette, vol. 21, no. 2, pp. 447–450, 2014.
5. Valent, O. - Galád, M. and Kačmár, Ľ.: Technická diagnostika 1, CMMS, Praha, Czech Republic, 2010.
6. Baron, P. - Kočiško, M. - Blaško, L. - Szentivanyi, P.: Verification of the operating condition of stationary industrial gearbox through analysis of dynamic signal, measured on the pinion bearing housing - 2017.In: Measurement. Vol. 96 (2017), p. 24–33. - ISSN 0263-
7. Poór, P. - Antal, P.: Modern maintenance management in a company on the principle of implementation of the computerized maintenance management system(cmms) maintenance assistant v3, Transfer of inovation 16/2010, Technical University of Košice, Košice, Slovakia, 2010
8. Kuruc, C.: Application of modern diagnostic methods for ecological optimalization of machines, Technical University in Brno, Brno 2006. 20 p.

Chapter 2
Technical Diagnostics

Technical diagnostics is currently one of the most important factors of reliability and thus the operability of machinery and mechanical equipment. The key attribute of a technical system is the reliability, defined as an ability to provide desired function while preserving the operation conditions in given time. This fact increases the maintenance demands on machinery, which has to ensure the intensive use of the means of production and equipment. Methods and means of technical diagnostics are often used to achieve this objective. Technical diagnostics is now considered an academic discipline, which tracks the manifestation forms of defects, development of methods, their detection, and principles of diagnostic devices operation [1].

The area of the interest of technical diagnostics is a technical system, which is an actual or expected subject for the verification of its technical state. Technical system may be a production unit, its subgroup, machine element, a separate product, etc. The basic task of diagnostics is therefore to provide the diagnosis characterizing the technical condition of the system in terms of occurrence of failures. Provided diagnosis must be usable for nursing actions optimization in order to bring the object to normal condition. The features of the diagnosed object are very important for the success of the diagnostic system. Among most important features, there is the capability to be diagnosed, which is expressed by the suitability of an object for the use of diagnostic devices. A machine is well diagnosable when all of the diagnostic actions can be performed easily with required accuracy and possibly at little cost [2, 3].

We verify the technical and operational state of the system, i.e., all the characteristics that represent its ability to perform required functions at a given moment under defined conditions of use. For the machinery products, this can be as follows [2, 4, 5]:

P. Baron et al., *Application of Troubleshooting Tools in the Monitored Production Processes*, Management and Industrial Engineering,
https://doi.org/10.1007/978-3-031-41428-2_2

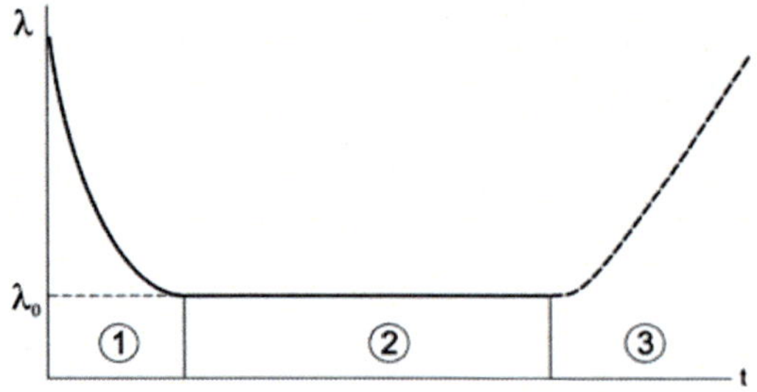

Fig. 2.1 Bathtub curve of failure rate dependence [1]. *1—area of frequent failures, 2—area where the level of failure rate is practically constant, and 3—area where the failure rate increases due to wear of parts*

- impeccable condition, when the object satisfies all the requirements set out by production-technical documentation,
- operable condition, when the object is able to perform the defined functions and follows the values of the given parameters within the limits specified by technical documentation,
- fault condition, at which the object is unable to perform the required function within the limits specified by technical documentation.

To increase a device reliability, it is necessary to meet the following requirements:

- increase the time in-between repairs (operation readiness),
- shorten the repair time,
- reduce the consequences of a breakdown by timely repairs.

Such requirements necessitate information on the actual condition of the equipment. This can also eliminate hidden, less serious defects, which can later develop into a larger, often fatal breakdown. Based on the above information, a malfunction can be addressed at the most appropriate time, both in terms of the nature of defect (the extent of the defect and the course it has run) and the impact on the production process (e.g.: the defect can be addressed at the weekend, during holidays, when the equipment is not involved in production). Proceeding in such way can eliminate not only the consequences, but also the initial symptoms of the malfunction and significantly increase the equipment's life [1].

Figure 2.1. depicts the relative failure of products in time. The above dependence describes the form of threat to the performance of the product function, which includes three parts:

- the first part is the decreasing failure rate, known as early failure.
- the second part is the constant failure rate, known as random failure.
- the third part is the increasing failure rate, known as wear.

Area 2 is referred to by the term of malfunctions intensity. In order to limit the impact of early malfunctions as much as possible, the so-called artificial aging of components is carried out. At the customer's request, the components are subjected to a defined load, a certain temperature and stress, and are subsequently all checked and sorted [1].

The aim of troubleshooting is to objectively know the technical condition of the monitored object and to determine its ability to perform the required functions under specified conditions, not only at present but also in the future.

Troubleshooting can be further characterized according to various aspects, e.g.: in terms of the task addressed, in terms of deployment, in terms of time management, in terms of the measured signal, etc.

Classification of troubleshooting in terms of the task addressed [1]:

- Troubleshooting without taking the equipment apart,
- Non-destructive troubleshooting.

Troubleshooting without the need for dismantling is further divided into:

- In terms of time:
  - cyclic (periodic),
  - permanent (monitored),
  - at the user's request.
- In terms of drawing judgment from the troubleshooting parameter measured:
  - single-parametric (one troubleshooting parameter addressed by a single troubleshooting method),
  - multi-parametric (a combination of troubleshooting parameters addressed by different troubleshooting methods); alternative multi-parametric (several troubleshooting parameters addressed by a single troubleshooting method),
- In terms of the task addressed:
  - detection of the existence of an emerging defect (detection),
  - establishment of the defective part or node (localization),
  - determination of the causes of emerging defects (specification),
  - prediction of residual life (prediction).

Detection, localization, specification, and prediction constitute four basic troubleshooting steps and cover the entire period of the final stage of the technical life of the respective node. They account for the most important and most intense phase in the machine's troubleshooting process [1, 6].

Methods of troubleshooting application:

Progressive troubleshooting

It is based on the analysis of the results of checks when the malfunctioning element is determined from combinations of different checks results. This method assumes that all checks are completed before their results are analyzed. The method is suitable for automatic check systems. The order of checks may be arbitrary, but the costs of individual checks are not interdependent. It is necessary to determine their optimal sequence so that the average cost of troubleshooting is minimal. The progressive method is not suitable for "manual" troubleshooting as it is lengthy due to requiring all checks to be performed [7].

Two-conditions troubleshooting

This method is still the most common way of assessing the technical condition of an object. Up to a certain level of troubleshooting signals, the element is in a normal condition, and after reaching this certain signal level, a diagnosis about a malfunction is made. This is a very simple way of troubleshooting. It is largely imperfect, mainly because it does not respect a significant practical requirement, which is forecasting of technical condition, especially when we are dealing with progressive malfunctions [7].

Multiconditional troubleshooting

There are several ways of applying this method. As a rule, a diagnosis is made from the level of the measured signal, where the nature of malfunction, expressed in quantitative terms, is assigned to a specified range of the measured values [7].

Active troubleshooting

This method involves special test signals the nature of which is to provide checking. There are four types of troubleshooting using vibrations and frequency analysis (troubleshooting from vibrations uses frequency analysis as a mathematical tool) [7]:

- comparative analysis,
- trending,
- analysis of malfunction sources and manifestations,
- analysis of malfunction causes.

Subjective troubleshooting

These methods do not make use of any measuring instruments, as they involve [8].

Checks done:

- visually,
- acoustically,
- thermally,
- aromatically.

Objective troubleshooting

Subjective methods are often insufficient to assess the actual condition of the object diagnosed. This problem is closely related to technical development and the use of more complex machines and equipment. A variety of devices allowing for assessment of a machine's technical condition are available. Various purpose-built checking and troubleshooting instruments and devices are used, operating on different principles [8].

Objective troubleshooting methods are divided into two basic groups [8]:

Functional methods—these are used where it is impossible to identify the physical processes that give rise to a malfunction. To meet the troubleshooting objectives, a direct comparison of the standardized outputs of the object are used with precisely defined inputs of troubleshooting parameters.

Physical methods—they monitor the incidental values of the production equipment that are not crucial for its operation, e.g., vibration, temperature, noise, etc.

Typical physical troubleshooting methods include, in particular [8, 9], the following:

- acoustic method dealing with the initial frequency spectrum (noise measurement),
- vibration method, using vibration measurement,
- ultrasound method, recording ultrasonic emission,
- thermal method, based on point temperature measurement or recording thermal fields,
- electromagnetic method, using a penetrating electromagnetic field,
- tribological method, using the knowledge of lubricants for troubleshooting purposes.

The input and output signals of troubleshooting instruments and equipment depend on the method used. Continuous monitoring and recording of the condition of the machine or its elements constitute an advantage [9].

PF scheme

PF scheme (Fig. 2.2) indicates the time of the malfunction inception and its impact on the system deterioration—we are talking about a potential failure. The first perception of defect dates back to this point. If no action is taken at this point, deterioration continues, mostly at an accelerated pace, until reaching the point (F)—a functional defect. The time between a potential and a functional defect is called the PF interval. It can be expressed in different units, related to the load rate (motor hours, number of products manufactured, number of cycles produced, etc.), but most frequently used is the unit of elapsed time. In general, the time between two measurements should be half of the PF interval. This time interval ensures that the troubleshooting prognosis detects a potential defect before a functional defect occurs and provides sufficient time to get ready for repairs [10].

In this case, ONLINE monitoring and locking systems are used, which shut down the equipment when the set values are exceeded, so as not to damage it.

PF scheme

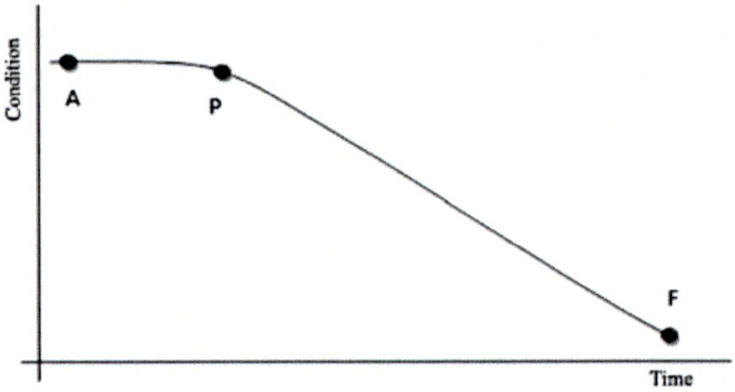

Fig. 2.2 PF scheme [10]. A—the point where the malfunction occurs, P—the point where we are able to perceive the malfunction for the first time, and F—the point where the malfunction will fundamentally affect the operation of the equipment

References

1. Janoušek, I. - Kozák, J. - Taraba, J.: Technická diagnostika. SNTL Praha, 1988. 432 p.
2. Baron, P. - Kočiško, M. - Dobránsky, J. - Pollák, M. - Cmorej, T.: Research and Correlation of Diagnostic Methods for Assessment of the State of Oil Filling in Cycloid Gearbox - 2015.In: Advances in Materials Science and Engineering. P. 97841–97841. - ISSN 1687-
3. Balog, J. - Chovanec, A. – Kianicová, M.: Technical Diagnostics, TnUAD, Trenčín, 2003.
4. KREIDL, M., ŠMÍD, R.: Technická diagnostika. BEN – technická literatúra, Praha. 2006. Vydaní 1. ISBN 80-7300-158-6.
5. Sekereš, J. – Turis, J.: Tribology as a Part of Machine Nodes Design, Scientific Studies, Technical University in Zvolen, Zvolen, 2009.
6. Valent, O. - Galád, M. and Kačmár, Ľ.: Technická diagnostika 1, CMMS, Praha, Czech Republic, 2010.
7. Zahradníček, R.: Vybrané problémy technickej diagnostiky a dôrazom na vibrodiagnostiku, Košice, 2001. ISBN 80-7166
8. Šenko, M.: Metódy technickej diagnostiky vo výbraných podnikoch a ich aplikácia. Bakalárska práca. 67 p. Prešov. 2013.
9. Ľubiščaková, A.: Zabezpečenie spoľahlivosti technickej diagnostiky. Novus Scientia, 2007.
10. Kováč, M.: Měření a analýza vibrací elektrického stroje. Bakalárska práca. Fakulta elektrotechniky a komunikačních technológií. Brno, 2009.

Chapter 3
Measurement and Assessment of Technical Systems' Vibrations

Vibrations represent an operating parameter enabling to assess low-frequency dynamic states, such as imbalances, misalignments, mechanical backlash stemming from the positioning, structural resonance, insufficiently rigid foundations, etc [1, 2]. Vibrations are incidental of the behavior of technical equipment during operation. In machines, vibrations arise from movement of their parts. In principle, they arise when the speed of the motion of the mass changes during rectilinear movement, movement along a curved trajectory, or rotational movement. Vibrations are also caused by mutual contact of individual parts. Since machines and equipment contain parts that move in such way, these are a source of forces causing vibration of the machinery as such. Each part of the machine is interconnected with a certain final stiffness with the final mass of individual parts. Such part forms a resonant mechanical circuit, which is excited by active excitation forces. Thus, each machine has individual vibration manifestations, and these are typical for how that particular machine works.

However, during operation of devices, the connections of the individual parts change. Bearings wear down, changes in the mass of worn parts occur, etc. These facts also result in a change in the vibration of the equipment in operation. As most malfunctions of rotary machines are manifested by excessive vibrations, vibration signals are used as indicators of the machine's mechanical condition. Each mechanical error or defect generates vibrations in its own specific way. In order to determine their cause and select the optimal remedial action, it is important to identify the type of vibration.

Vibration measurement is not a major problem. The aim is to analyze the signals and subsequently identify the causes of the problem. In the analysis of vibration signals, attention is paid to its two basic components: frequency and amplitude. Frequency represents the number of occurrences of a certain phenomenon over a certain period of time. Based on the vibration frequency, it is possible to characterize a type of malfunction. Identification of the frequency at which the vibrations occur

P. Baron et al., *Application of Troubleshooting Tools in the Monitored Production Processes*, Management and Industrial Engineering,
https://doi.org/10.1007/978-3-031-41428-2_3

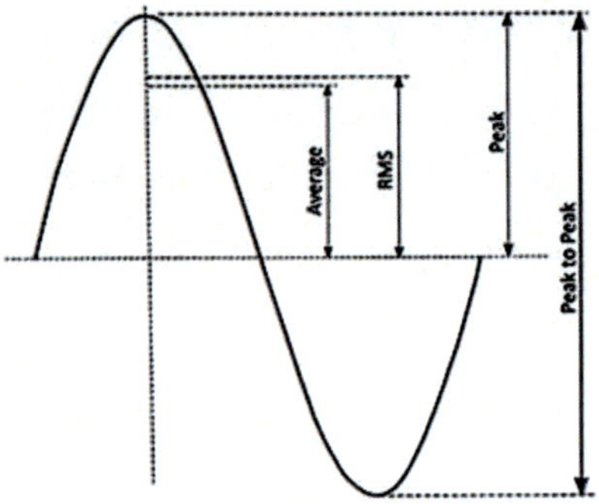

Fig. 3.1 Total vibrations parameters

paints a picture of their cause. The amplitude represents the magnitude of the vibration signal. It is related to severity of the malfunction [1, 2].

Total vibrations represent the total vibrational energy measured within a certain frequency range. The following expressions are used to measure total vibrations (Fig. 3.1) [1, 2]:

- peak value (PV)
- peak to peak (PtP)
- average value (AV)
- root mean square amplitude (RMS)

An important method for monitoring vibration signals and their subsequent analysis is the FFT method—fast Fourier transformation. Through the FFT vibration analysis, it is possible to identify whether a device malfunctions and what problem in the device is the source of such excessive vibrations. Since it is clear that a certain defect occurs at certain frequencies, the FFT spectra are analyzed through monitoring changes in amplitudes in these frequency ranges [1, 2].

3.1 Balancing of Rotating Parts of Machines

3.1.1 Machine Imbalance

Machine rotor imbalance is understood as the condition under which the residual imbalance or oscillation of the bearing studs at operating speeds is outside the specified limits. Its main central axis of inertia is generally not identical with the rotor's axis of rotation.

Imbalance occurs due to inaccuracies in the manufacture and assembly of rotor parts or it occurs during operation.

Unbalanced rotor parts are the source of centrifugal, rotor-speed driving forces that are demonstrated as vibrations and noise.

Rotor imbalance causes the following:

- increased dynamic strain of the rotor,
- reduced bearing life,
- may cause the rotor parts to touch the stator,
- causes the machine anchor system to loosen or get damaged.

There are four types of imbalance:

1. static imbalance—the main central axis of the rotor inertia is parallel to the rotor axis,
2. quasi static imbalance—the main central axis of the rotor inertia is independent of the rotor axis and does not pass through its center of gravity,
3. pair imbalance—the main central axis of the rotor inertia is independent of the rotor axis and passes through its center of gravity,
4. dynamic imbalance—the main central axis of the rotor inertia is offset with respect to the rotor axis [3, 4].

3.1.2 Diagnostic Symptoms of Imbalance

Imbalance of rotating parts of the machine is relatively easy to diagnose, but some diagnostic symptoms are different for each type of imbalance. Basic common diagnostic symptoms of imbalance:

- in the frequency spectrum of radial vibrations, vibrations at the rotational frequency are dominant,
- the amplitude of the radial vibration vector at the rotational frequency is approximately the same in the horizontal and vertical plane,
- the phase shift between the radial vibration vectors at the rotational frequency in the horizontal and vertical planes is approximately 90° (±30°),
- the amplitude of the vibrations increases approximately with the second power of revolutions,
- the phase difference of radial vibration vectors at the rotational frequency on both bearings of the rigid rotor machine is approximately the same (±30°) in the horizontal and vertical planes.
- the phase difference of radial vibration vectors at the rotational frequency on the inner (outer) bearings (relative to the coupling) of the device is approximately the same (±30°) in the horizontal and vertical plane.

The following diagnostic symptoms are yet typical for each type of imbalance:

(a) static imbalance

- vibrations in the radial direction are substantially higher than in the axial direction,
- the phase difference of radial vibration vectors at the rotational frequency on both bearings of the machine is approximately equal to 0° (±30°) in the horizontal and vertical planes.

(b) pair imbalance

- the phase difference of radial vibration vectors at the rotational frequency on both bearings of the machine is approximately 180° (±30°) in the horizontal and vertical planes,
- significant pair imbalance sometimes generates high vibrations in the axial direction [3, 4].

3.1.3 General Principles for Rotor Balancing

When balancing the rotor, it is desirable to minimize dynamic forces transmitted to the machine from its rotor by aligning the main central axis of the rotor inertia with its axis of rotation. This is accomplished by adding, removing, or displacing the correction matter in one, two or more balancing planes.

According to the number of balancing planes and the possibility of eliminating a certain type of imbalance, two basic balancing procedures are distinguished:

- static balancing—balancing in one plane,
- dynamic balancing—balancing in two or more planes—all kinds of imbalances can be eliminated [3].

According to the rotor balancing method, three balancing modes are distinguished:

- balancing on gravity balancing devices by weighing or swinging the rotor. Only static rotor balancing can be performed in this way.
- balancing on centrifugal balancing machines during rotation. Static and dynamic rotor balancing can be performed in this way.
- operational balancing in its own frame and its own bearings. Static and dynamic rotor balancing can be performed in this way.

During operational balancing, machine vibrations are measured, such as its response to dynamic forces transmitted from the rotor. In measuring these vibrations, the amplitude and vibration phase of the vibration vector are generally measured, while it is not critical which vibration variable (deflection, velocity, acceleration) is measured.

The result of the operational rotor balancing process is relatively largely influenced by the size and location of the test matter during the test run of the machine. In

Table 3.1 Recommended alignment tolerances [3, 5]

	Angular misalignment of axes [mm/100 mm]		Parallel misalignment of axes [mm]	
rpm	Optimal value	Limit value	Optimal value	Limit value
0–1000	0.06	0.10	0.07	0.13
1000–2000	0.05	0.08	0.05	0.10
2000–3000	0.04	0.07	0.03	0.07
3000–4000	0.03	0.06	0.02	0.04
4000–5000	0.02	0.05	0.01	0.03
5000–6000	0.01	0.04	>0.01	>0.03

determining the weight of the test matter, it is assumed that its mass should cause a centrifugal force equal to one tenth of the force of strain acting from the rotor to the bearing. When placing the test matter in the balancing plane, it is desirable that the matter is not placed into the "hard" point of the rotor. However, establishing this point is a problem, as normal measurements fail to identify it unequivocally.

The final criterion of the correctness of the weight of the test matter and its placement is, in particular, a sufficient change in the phase of the vibration vector in the next test run of the machine. This change should be at least 25°–30°. Changing the vibration amplitude is not so critical but changing it by at least 30% will increase the quality of the balancing process [3, 4].

Under real-world operating conditions, the alignment tolerance for rotating machines depends on a number of factors (operating speed, efficiency, clutch type, length of the inserted shaft, life of the technical equipment, type of equipment). Under practical circumstances, it is problematic to consider all the influences—a certain simplification is applied, expressed in the form of recommended values of alignment tolerances (Table 3.1)

3.1.4 Operational Balancing Methods

1. Methods Without Phase Measurement

This method does not measure the phase, but overall vibrations. Here, a three-position balancing method and a multi-position balancing method that we can use for manual measurement without a balancer are applied [3, 4].

Multi-position balancing method—this method requires that the rotor circumference in the balancing plane be divided into at least five equal parts (preferably 8 parts). Before balancing, the vibration of the bearing stand must be measured. To calculate the weight of the auxiliary load, the following relationship is applied (3.1):

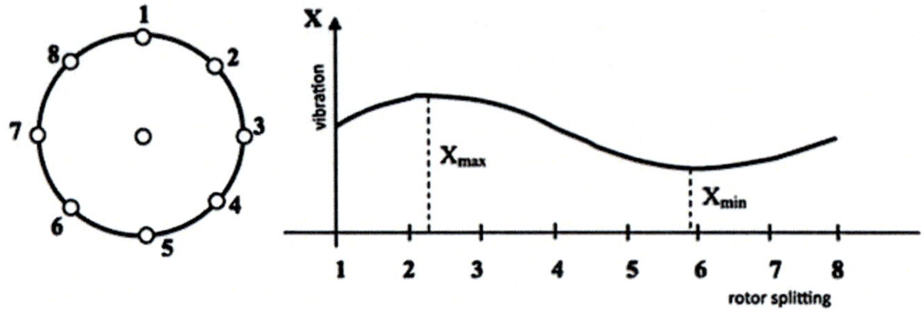

Fig. 3.2 A multipositional balancing method graph [3, 4]

$$m_p = \frac{m.|X_A|}{r_p} \tag{3.1}$$

where:

m_p weight of the auxiliary load [kg 10^{-3}]
m weight of the rotating part (impeller, shaft, clutch)
X_A the vibration velocity amplitude (RMS)
r_p auxiliary load radius—distance of its center of gravity from axis of rotation

After calculating the weight of the load, it is placed in all positions, and the vibration of the bearing stand is then measured. A graph of dependence of the measured kinematic quantity on the load position is made (Fig. 3.2). The position of the final load is subtracted from the graph and its weight is determined by calculation. The final load is placed on the rotor and the test is performed [3, 4].

In case the relationship (3.2) applies:

$$X_A \geq \frac{1}{2}(X_{\max} + X_{\min}) \tag{3.2}$$

The weight of the final load is calculated according to the relationship (3.3).

$$m_v = \frac{X_{\max} + X_{\min}}{X_{\max} - X_{\min}} \tag{3.3}$$

Otherwise, if:

$$X_A < \frac{1}{2}(X_{max} + X_{min}) \tag{3.4}$$

The weight of the final load is calculated according to the relationship (3.5).

$$m_v = \frac{X_{\max} - X_{\min}}{X_{\max} + X_{\min}} \tag{3.5}$$

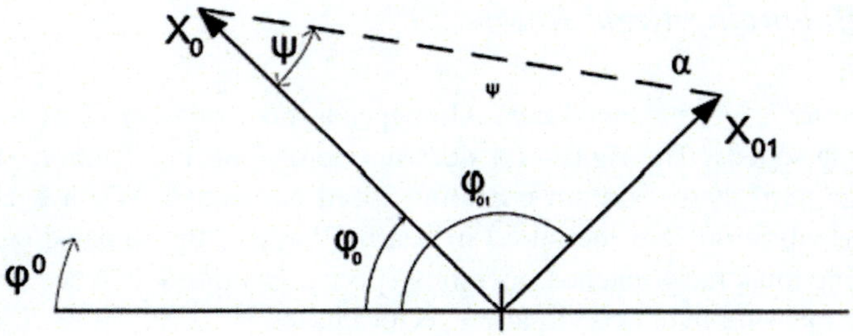

Fig. 3.3 Principle of a single-positional method

2. Balancing Methods with Phase Measurement

Single-positional balancing method is the simplest method compared to other balancing methods. In this method, the oscillation phase must be measured (optical marks and optical probes are used, or the rotor is illuminated with a stroboscopic lamp synchronized with the non-weight balance vector).

When the rotor is spun, the amplitudes and the oscillation phase of the bearing stand are measured. The amplitude X_0 and the phase ϕ_0 of the oscillation deflections are measured at point A (Fig. 3.3). The test load is then placed in a selected position of m_p weight and at the same revolutions as in the previous case, the amplitude X_{01} and the oscillation phase ϕ_{01} of the oscillation deflections are measured at point A.

The evaluation of the measurement is performed by plotting the oscillation vectors from the two measurements in a polar diagram at the selected scale (Fig. 3.2). Difference of vectors X_{01}–X_0 determines the vector coefficient α. The angle ψ, by which this vector had to rotate to merge with the vector X_0, but with an opposite meaning, indicates the position of the desired load on the rotor relative to the position where the auxiliary load was added. The weight of the desired load is established from the relationship (3.6).

$$m_v = \frac{X_0}{\alpha} m_p \frac{r_p}{r_v} \tag{3.6}$$

where

m_v the final load weight
m_p the test load weight
X_0 amplitude of the oscillation deflection at point A
α vector coefficient
r_p auxiliary load radius—distance of its center of gravity from axis of rotation
r_v the final load radius—the distance of its center of gravity from the axis of rotation

After the weight is calculated, the load is placed at the specified position and the test run of the rotor is performed [3, 4, 6].

3.1.5 Balancing Rigid Rotors

The rigid rotor can be balanced even at low speeds, but it does not need to be balanced at operating speeds. The rigid rotor does not deform during rotation, its axis does not change until the maximum operating speed is reached. When balancing such rotors, the vibrations are measured in bearing bays in the radial direction where the machine must have reached operating speeds. Operating balance procedure for amplitude measurement and vibration vector phase:

1. Basic machine run and amplitude measurement and vibration vector phase at rpm,
2. placement of test load,
3. test run of the machine and measurement of the amplitude and phase of the vibration vector at the rotational frequency,
4. establishment of the load size and placement through a graph or through computation,
5. test run of the machine with test load,
6. if the imbalance is still large, the size and placement of the additional balancing matter is calculated, but the original load remains in place [3, 6].

In general, the greater the weight of the rotor, the greater the imbalance under the same conditions.

In calculating the permissible residual imbalance, we draw on the Eq. (3.7).

$$\varepsilon_p = \frac{N_p}{M} \tag{3.7}$$

where

ε_p specific residual imbalance [μm]
N_p permissible load [g mm]
M weight of the balanced tool [kg]

An important parameter in determining the permissible imbalance is the operating speed of the rotor. The degree of admissible imbalance is calculated from Eq. (3.8).

$$Q = \varepsilon_p \cdot \omega \cdot 10^{-3} = const. \tag{3.8}$$

where

Q permissible residual imbalance [mm s^{-1}]
ε_p specific residual imbalance [μm]
ω angular velocity [rad s^{-1}]

Eccentricity of imbalance is calculated from Eq. (3.9), that is, when the value decreases, the center of gravity approaches the axis of rotation [3, 5].

$$e = \frac{U}{M} = \frac{m.r}{M} \tag{3.9}$$

where

e permissible residual eccentricity [μm]
U permissible residual imbalance [g mm]
M weight of the balanced tool [kg]
m permissible residual weight of the imbalance [g]
r position radius of residual imbalance [mm]

3.1.6 Analysis of the Operating Condition of the Furnace Exhaust Fan Depending on Its Impeller Alignment

Based on the requirement of our partner in the metallurgical industry, we performed a balancing of the furnace exhaust fan impeller with a diameter of 1.5 m. The impeller is fitted with eight 20-cm-wide blades. The fan is connected by means of a disc clutch to an electric motor with a power of 45 kW and a speed of 1500 rpm (Figs. 3.4 and 3.7) (Figs. 3.5 and 3.6).

The measurement was carried out as follows:

Dynamic balancing of the fan impeller was performed using the Microlog CMVA 10 frequency analyzer. The assessment of the fan's operating condition was carried out by measuring and evaluating vibration frequency spectra. In the test operation, vibrations were measured on both bearings (Fig. 3.4) including the phase. Since the measured vibration value was higher than recommended, it was necessary to balance the machine.

Fig. 3.4 Placement of the emissions fan

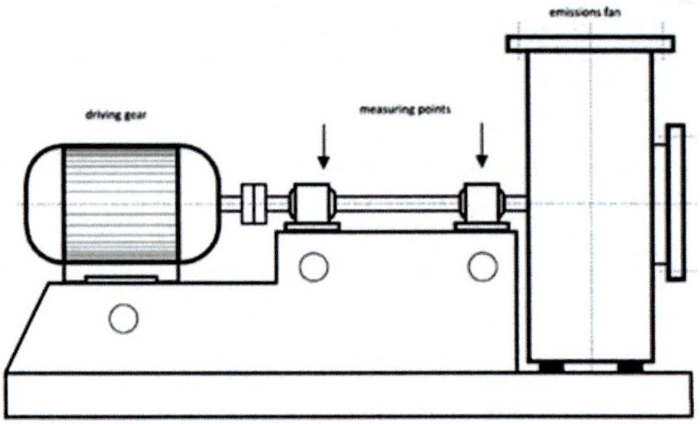

Fig. 3.5 Scheme of the monitored device with the location of the measuring points

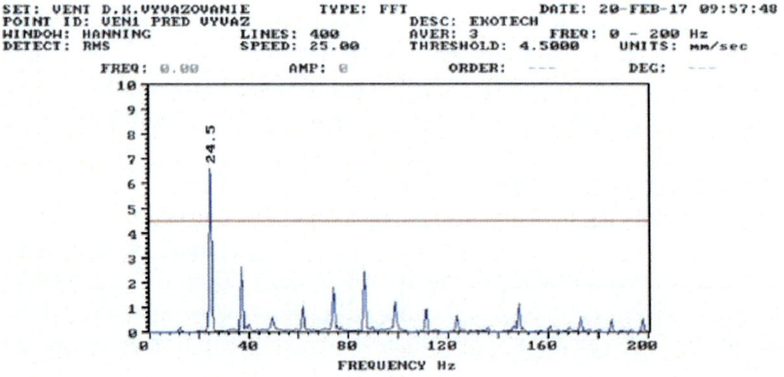

Fig. 3.6 Condition (velocity) before balancing the fan

Table 3.2 lists the recommended values for mechanical vibration measurements, in accordance with the ISO 10816-3 standard.

Figure 3.6 shows the mechanical vibrations measured before balancing at a rotational frequency of 24.5 Hz. Looking at the graph, it is possible to determine the impeller imbalance, which is at 6.9 mm s^{-1}.

The impeller circumference has been divided into four equal parts with an angle of 90° in the direction of rotation of the fan, which is shown in (Fig. 3.7). The first point was selected at the 30° position, where the test weights were welded.

Table 3.2 Recommended vibration values

Measuring method	ALARM 1 (Warning)	ALARM 2 (Danger)
MFV [mm/s]	4.5	11.2
EN3 [Eg]	4.5	11.2

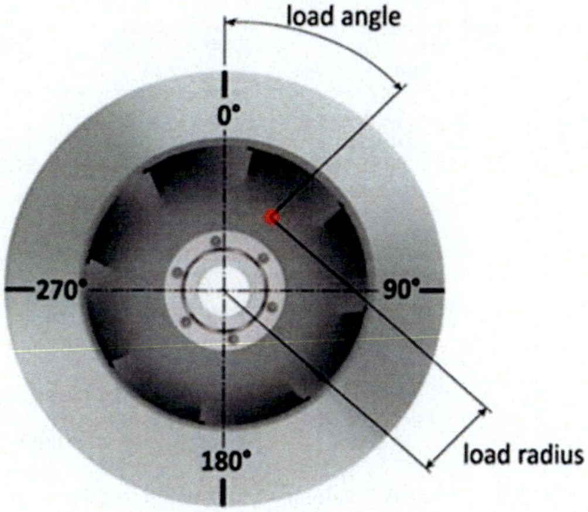

Fig. 3.7 Splitting the rotor circumference and placing the load

After the load was attached by welding (Fig. 3.8), the fan was spun to operating speeds, and vibration and phase measurements were performed on both bearings. Applying the Microlog CMVA 10 analyzer, we calculated the weight of the test load and the location of its placement. The test balance load was removed with a grinder and adjusted to meet the required weight according to the respective calculation. In the next step, the load with required weight was attached by welding to the fan's circumference. A follow-up fan run was performed, in which vibrations and phases were re-measured on both bearings. The measurement results confirmed the correctness of the fan impeller balancing procedure. The resulting impeller imbalance is 2.0 mm s^{-1}. The measurement results are within the limits of the recommended values given in the ISO 10816-3 (Fig. 3.9) standard.

Table 3.3 shows balance load weights and position of their placement. The balance loads made of the following material were used—the structural steel EN ISO 295.

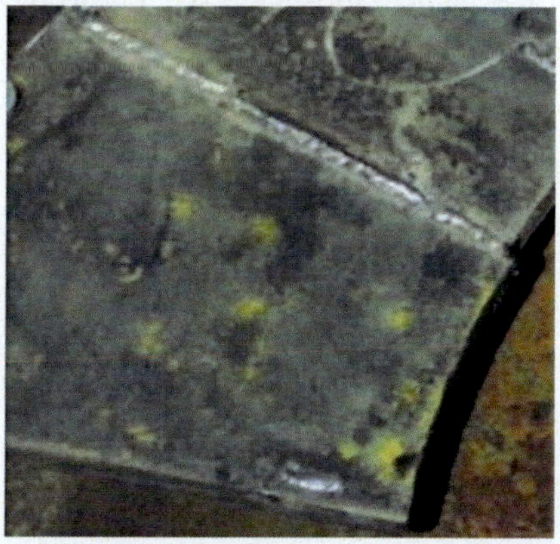

Fig. 3.8 Applied balance load

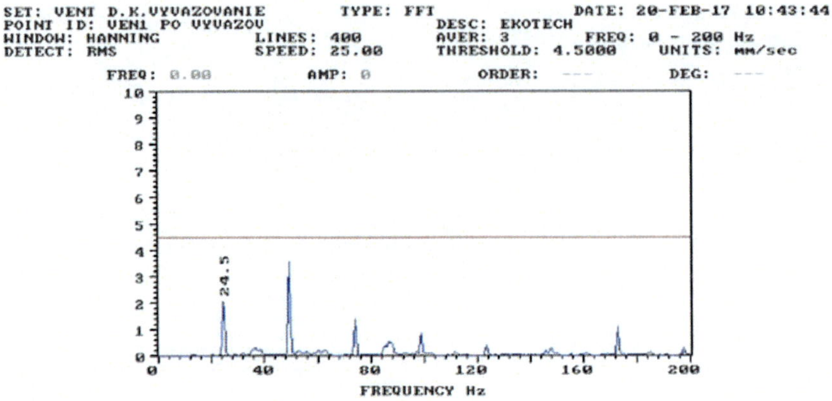

Fig. 3.9 Condition (velocity) after balancing the impeller fan

Table 3.3 Test balance load weight

Measurement number	Balance load weight [g]	Angular position [°]
1	50	30
2	30	25

The proper machine balancing was successful. The vibration values on both bearings were below the specified limit (Table 3.4). The final load was firmly welded so that it does not loosen during operation, which would cause the fan impeller to resume its state of imbalance. Figure 3.10 shows a fan with the final balance load attached by welding.

Table 3.5 shows an evaluation of the operating state of the nodes of the fan measured after balancing.

At the end of the balancing process, measurements were made to determine the mechanical loosening (backlash) of the bearings at the coupling and the impeller. Figures 3.11 and 3.12 show the amplitude of the vibration signal—spectrum 3 of the envelope. It is obvious from the record that the bearings are without significant damage, but with increased radial backlash.

Table 3.4 Vibration values measured before and after balancing the fan impeller

Fan	Mechanical vibrations at 24.5 Hz (IMBALANCE) MFV [mm/s]	Total mechanical vibration MFV [mm/s]
Impeller before balancing	6.9	8.4
Impeller after balancing	2.0	4.3

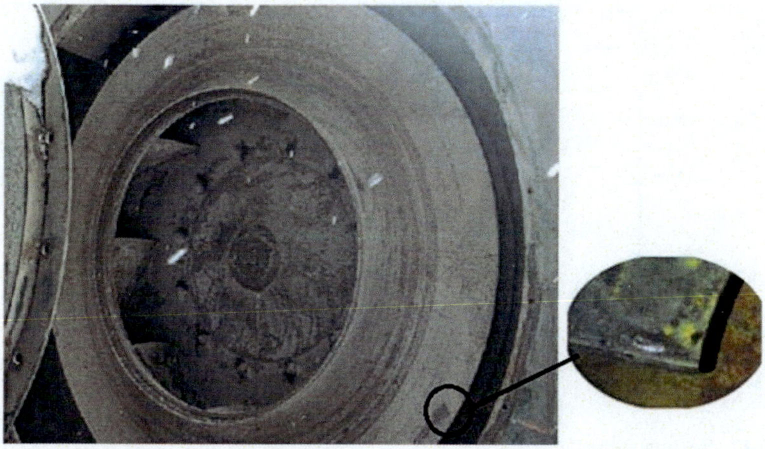

Fig. 3.10 Aligned fan and the load detail

Table 3.5 Evaluation of the operating state of the monitored fan nodes

Machine and measuring node name	Operating state	Note
Fan—engine	Satisfactory	Bearings OK, mechanical vibrations in the axial direction slightly increased
Fan—bearings	Satisfactory	Mechanical vibrations—bearings without significant damage, but with increased backlash

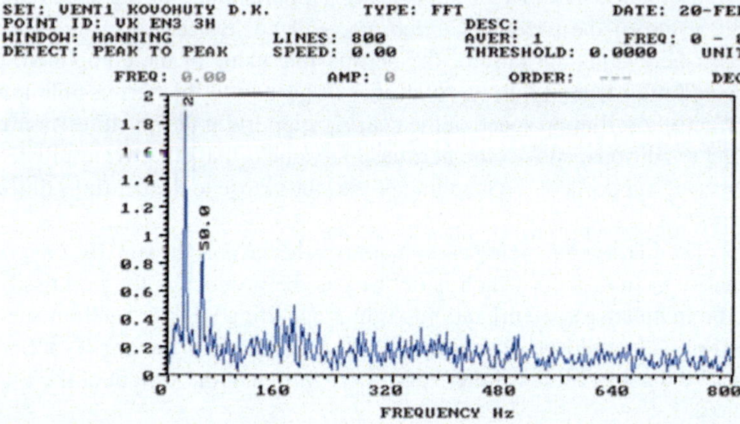

Fig. 3.11 Bearing at coupling

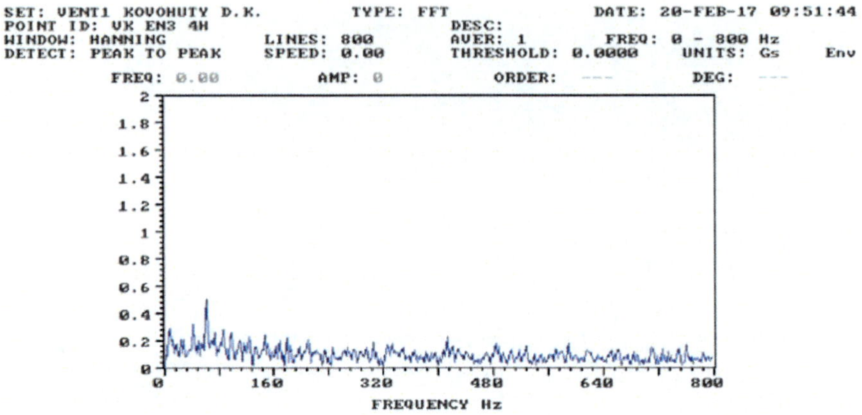

Fig. 3.12 Bearing at the fan

3.1.6.1 Standards for Rotor Balancing—Discussion

The results of the operational rotor balancing or balancing of a rotor assembly are evaluated according to the maximum value of the oscillating vibration variable measured at all machine bearing supports in the vertical and horizontal directions perpendicular to the rotor axis. These measurements are performed over the full speed range, idling, and rated load.

The balancing of the rotor is considered to be completed if the maximum value of the effective oscillation velocity of the bearing supports at the idling speed does not exceed the value of the product C_0 *and* V_e, will not exceed the value V_e at critical speeds, where C_0 is the ratio of the permissible value of the component of the effective oscillation velocity at the rotational frequency to the permissible total value of the effective oscillation speed of the bearing supports at the machine running under operating conditions, and V_e the permissible value of the effective oscillation speed of the bearing supports of the machine or the studs close to the bearings during operation [mm s^{-1}].

The quality of solid rotor balancing is evaluated on the basis of the so-called degree of balance quality Q [mm s^{-1}], which is defined as the product of the permissible residual specific imbalance ε_p [μm] and maximum operating angular rotation speed of the rotor ω [rad s^{-1}], while the permissible residual specific imbalance ε_p is defined as the portion of the permissible balance load N_p [g mm] and the weight of the rotor M [kg] balanced.

Degrees of balance quality *Q* are divided into 11 classes, which then include individual types of rotating machines. Class Q0.4 *contains* machines with the highest quality value of the balance (0.16–0.40 mm s^{-1}), and class *Q4000* contains machines with the lowest quality value of balance (1600–4000 mm s^{-1}).

When balancing the rotors, it is also appropriate to determine the so-called rotor sensitivity to imbalance. This parameter is easy to establish from a graph of the results of the rotor test run at operational balancing, since the sensitivity of the rotor to the imbalance is expressed as the product of the test compound test matter and the radius of its placement to the amplitude of the rotor vibration vector at the rotational frequency caused by the addition of the test matter to the rotor.

The residual imbalance of the rotor U_z [g mm] after balancing has been performed is then determined as the product of the rotor sensitivity to imbalance and the amplitude values of the vibration vector at the rotational frequency after rotor balancing [3, 4].

3.1.6.2 Verification of the Operating State of the Monitored Fan After Its Impeller Has Been Balanced

Fifteen months after the process of balancing the emissions fan impeller of an industrial furnace, our team conducted vibration measurements on the device in question to verify its current technical condition.

The total vibrations of the device in the low-frequency and high-frequency range were measured, with the measurement points according to Fig. 3.13.

The purpose of the measurement was to investigate the total vibrations of the device. In particular, mechanical conditions, such as potential imbalance, axial misalignment, mechanical loosening, bent shaft, resonance, transmission problems, etc., were investigated. These measurements are defined in the standards dealing with permissible oscillation magnitudes on given devices (EN 122,011—Fans, EN 105,041—Compressors, EN ISO 10816—general standard for most machinery). The Microlog CMVA 10 Frequency Analyzer and accelerometer sensors of 100 mV/g were used to experimentally identify and analyze the oscillatory motion of the monitored device.

Fig. 3.13 Location of the measuring points on the fan device

Table 3.6 Alarm levels for LF vibration

Alarm A1—warning	4.5
Alarm A2–danger	11.2

Table 3.7 Alarm levels for HF vibration

Alarm A1—warning	7.1
Alarm A2–danger	11.2

To assess vibration in the low frequency range, the following were used:

- Velocity measurement method (jitter speed), [mm s^{-1}], RMS detection
- pursuant to recommendations of ISO10816-3, the FFT spectrum for the frequency range 2–1000 Hz
- Enveloping Acceleration method of measurement, [gE], Peak detection, for the frequency range up to 1000 Hz, FFT spectrum, and Time recording

For the assessment of vibration in the high frequency range, the following was used:

- Acceleration (jitter acceleration) method of measurement, [g], PtP detection, frequency range up to 6.4 kHz, FFT spectrum, and Time recording
- Enveloping acceleration method of measurement, [gE], peak detection, frequency range up to 10 kHz and up to 20 kHz, FFT spectrum, and time recording

From the point of view of the total vibration values measured, it was necessary to identify the recommended limits with respect to the requirements of the technical standard STN ISO 10816-3 [3].

For various machine operating modes (with respect to speed, load, power), the alarm limits (alarms) can also be determined on the basis of a general recommendation (200% signal increase over the reference value)—Tables 3.6 and 3.7.

Table 3.8 contains measured vibration values.

3.1.6.3 Validation and Evaluation of Measured Data

The values measured on the engine are low in all parameters (up to 1 mm s^{-1} or up to 1 g). LF vibrations on the impeller do not exceed the alarm value A1 (warning) either. Values measured using HF methods, however, point to a dissatisfactory operation. All values exceed A1 alarm (warning), some even alarm A2 (danger)—see. Table 3.8. Clear resonance regions are visible in acceleration spectra in the area around 3 kHz. Spectra of the 3rd envelope do not contain any dominant frequency (Figs. 3.14, 3.15, 3.16 and 3.17).

Table 3.8 Vibration values measured at defined measurement points

	Vel (mm/s)	Acc (g)	En3 (gE)
Point 1H	0.70	0.41	0.55
Point 1V	0.76	–	0.49
Point 2H	0.59	0.52	0.61
Point 2V	0.82	–	0.53
Point 2A	0.31	–	–
Point 3H	1.69	**11.51**	**16.49**
Point 3V	1.00	*11.17*	**16.28**
Point 3A	0.65	–	–
Point 4H	1.83	*7.68*	*9.09*
Point 4V	1.45	*9.48*	**12.01**

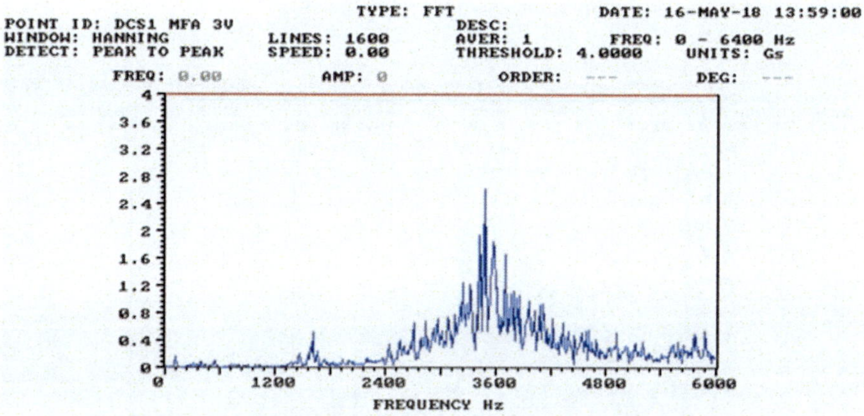

Fig. 3.14 Jitter acceleration spectrum (Acc) at point 3 V

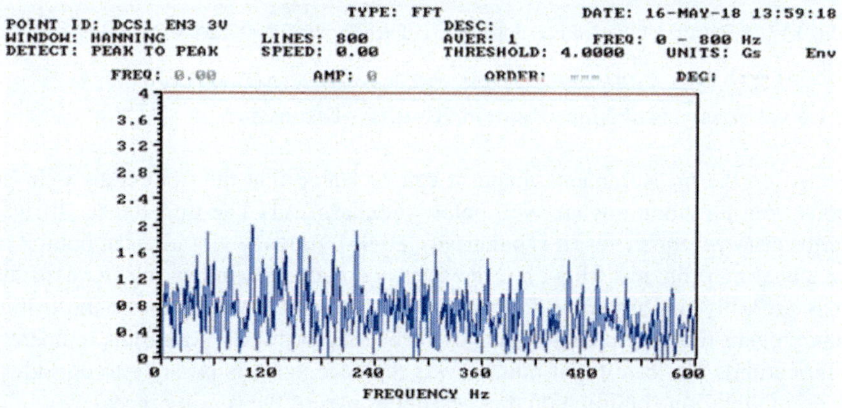

Fig. 3.15 Spectrum of the 3rd envelope (En3) of acceleration at point 3 V

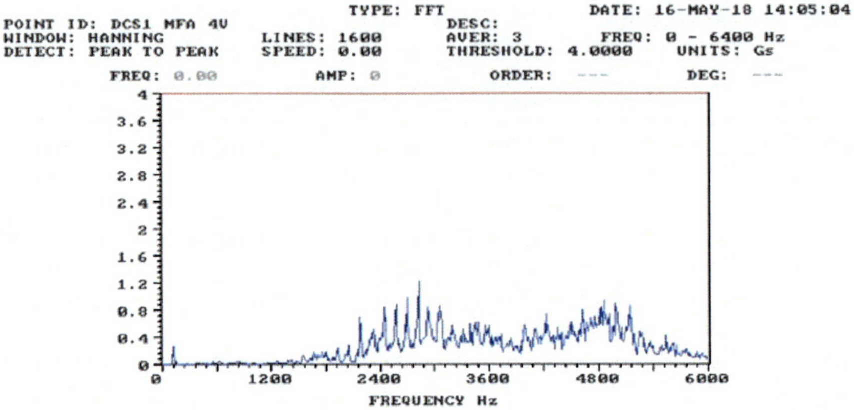

Fig. 3.16 Jitter acceleration spectrum (Acc) at point 4 V

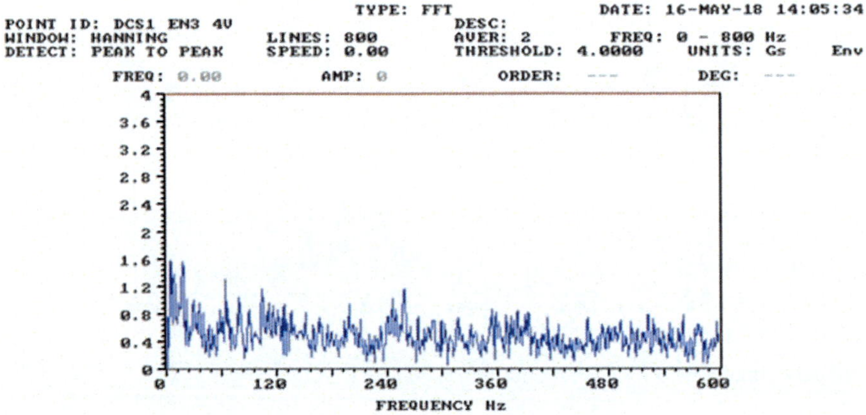

Fig. 3.17 Spectrum of the 3rd Envelope (En3) of acceleration at point 4 V

3.1.6.4 Analysis of Measurement Results—Discussion

Based on the measurements made, it can be stated that the fan engine is in good condition, its vibrations are well below the alarm A1. The impeller is all right in terms of low frequencies (it is balanced). High-frequency vibrations at point 3 indicate bearing problems. However, the spectra do not contain bearing frequencies, so it is probably not a bearing failure. We recommend the operator of the technical equipment a more frequent tracking of the bearing in point 3 (vibration, temperature, lubrication). The bearing at point 4 was replaced, so it is possible to consider the present vibration condition on this bearing as part of the run-in process.

Dr. Jerzy Piotrowski, in his published works, states that up to 99% of technical equipment is working with some axial misalignment. That is to say, an absolutely perfect alignment is not possible. Slight misalignment and low vibrations are, actually, often desirable to allow for functional lubrication, which is often done hydrodynamically [3, 7]. The most important goal of alignment is to increase reliability and extend the life of the technical equipment. Efforts aiming at accurate alignment are done with the following partial goals [3, 7]:

- reducing excessive axial and radial forces in the bearing, resulting in increased lifetime and dynamic stability of the rotor,
- avoidance of damage to the shaft due to material fatigue caused by cyclical forces,
- minimizing wear of the coupling part,
- minimize shaft deflection between power transmission point (for clutch) and end bearing,
- reduction of energy consumption (the examples shown show a saving of 2 ÷ 17%),
- reducing the vibration level of machine covers, bearing stands, and rotors.

Based on the research in the field of vibrational diagnostics and rotor balancing, we can confirm the fact that the loss of balance occurring during rotation is due to uneven weight distribution that creates a centrifugal force manifested by vibrations. This centrifugal force grows quadratically with operating speed. This means that at the same loss of balance, for example at the operating speed of 16,000 min^{-1}, the centrifugal force affecting the rotor is 16× greater than at the operating speed of 4,000 min^{-1}.

During rotation, the unbalanced rotor tends to rotate around its central axis of mass. As the bearings limit this movement, the centrifugal force causes the rotor to vibrate due to loss of balance. These vibrations primarily cause damage to the bearings, generate excessive noise and, in extreme cases, even lead to disintegration of the entire mechanism. Therefore, it is essential to minimize the loss of balance to an acceptable level.

Of course, development of modern production technologies has minimized the need for balancing in low-speed applications. On the other hand, dynamic balancing becomes a necessity in the application of rotary machines with growing operating speed.

Based on the analysis and monitoring of the technical system under evaluation—namely the industrial furnace fan, the following general conclusions can be stated, equally applicable to the new rotary equipment commissioned:

- Although we assume that each component forming the rotary equipment was well balanced, 100% balance does not exist.
- Although each rotating unit (engine, clutch, rotor, fan impeller) meets the maximum criteria of residual loss of balance, the resultant loss of balance of the overall fan device driven by the electric motor may be unsatisfactory when the assembly has been completed.
- Under real conditions, the position and extent of the displacement of the center of gravity from the axis of rotation is unknown (the relative position of the rotating

units is random after assembly). It is thus quite natural that the control measurements applied can identify with great probability higher vibration values than those recommended by the technical standards.

- The above-mentioned deficiency in both, the monitored and the new rotation systems, must be removed by additional balancing.

From the rotary machines operation point of view, it is argued that the main problem of their lost balance is the occurrence of vibrations, which not only cause noise, but also an increase in mechanical stress. Such stress is most common in bearings, where it reduces their service life. The loss of balance is closely related to greater weight of the machines, and this is the reason why balancing is a necessity, especially when it comes to heavy rotating components, including the impeller of the industrial furnace fan.

This analysis described the sequence of steps performed in balancing the emissions fan impeller of an industrial furnace. After the addition of a balance load with weight established by calculations, a test operation of the fan was performed, with measuring vibrations and phases on both of the monitored bearings of the fan rotor. The resulting impeller imbalance reached 2.0 mm s^{-1}. Thus, the results of the final measurement were within the limits of the recommended values prescribed by the ISO 10816-3 standard. Verification of the technical condition of the fan rotor bearings has not shown any significant damage. Fifteen months later, upon request of our partner, we verified the actual technical condition of the emissions fan in question, with the application of vibro-diagnostics tools. Based on the measurements made, it can be stated that the fan engine's condition is satisfactory. The impeller is all right in terms of low frequencies—it is balanced. High-frequency vibrations at measurement point 3 identified some bearing problems. However, the spectra did not contain bearing frequencies, which led to the conclusion that there was no bearing failure. We recommended the operator of the installation to track the bearing at point 3.

Our analyses, too, confirmed the connection between a reliable operation of rotary machines and the precise alignment of their rotors. Up to 50% of rotary machines failure is due to the incorrect alignment of their rotors. Device operators often rely on the clutches to bear considerable misalignment. Sometimes all the way up to 10 mm of radial retraction. However, the strain on the shafts, bearings, and seals increases dramatically as a result of the increase in the alternating forces and loads [3, 8, 9].

3.2 Diagnostic of Rolling Bearings

Technical diagnostic is applying different methods to the needs of machine diagnostics. It often happens that measurements are not interpreted, resp. the interpretation is not correct. For example, during the vibration diagnostics, the factor of creation of forces and energies that create fatigue damage and shorten the lifespan of the machines and machine parts is often forgotten. It means that twice more force shortens the life of the bearings on 1/8 and product life is shortened from five to ten years to a few months [2].

Diagnostic of bearings is a very difficult area. Bearing damage can be caused by unsuitable lubricant, contamination of lubricant, water, excessive force, oxidation of lubricants or by degradation of additives. In this case, it is necessary to think about the influence of the human factor. The bearing may be installed with a larger overhang, or was overheated during the assembly process, greased, resp. non-greased. All these factors can shorten the life of the technical system for a few hours to several months [2].

To ensure the quality of maintenance of rotating machines it is essential to interprete and evaluate the causes of the failures correctly. It happens quite often that a cooperation between maintenance, technical diagnostic and functional information processing for evidence of repairs, lubrication plans, and dates of technical diagnostics is missing.

3.2.1 *Diagnostics of Roller Bearings—Crest Factor*

Crest factor coefficient represents the basic methodology of diagnosis of rolling bearings. The method is based on measuring of effective and high vibration levels and on calculating their ratio PEAK/RMS (Fig. 3.18).

Because the ratio is evaluated by two values, this method depends on the type of bearing and on the shaft speed.

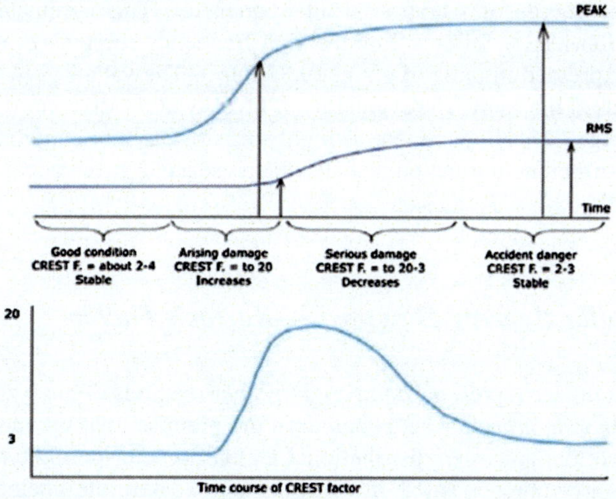

Fig. 3.18 Crest factor

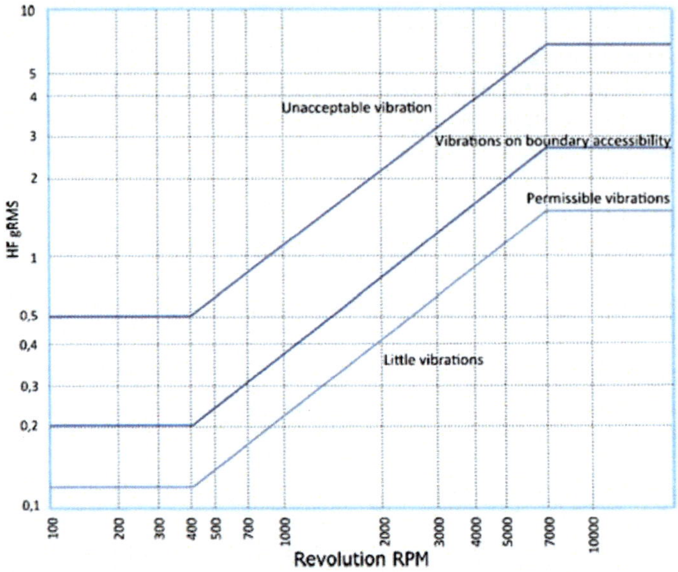

Fig. 3.19 Assess the severity of the fault applying method HF

3.2.2 *Diagnostics of Roller Bearings—HF (High Frequency Emission)*

Measuring of HF parameter is based on the recognition that the early damage increases the energy of vibration at all frequencies. This parameter is also very sensitive to lubrication failures. Emitted high-frequency energy is evaluated in RMS value and expressed in units of grams (9.81 m/s^2). The values g_{RMS} are dependent on the rounds of the shaft. The evaluation of severity of the disorder is applied on the scheme of Fig. 3.19. HF parameter is highly sensitive to lubrication disturbances. It indicates insufficient lubrication, but also the presence of mechanical contamination of the oil [10].

3.2.3 *Roller Bearing Diagnosis—Kurtosis Factor*

Kurtosis parameter represents compared to other methods of analyzing the state of bearing a static approach. This is based on the premise, that the random vibration signal has normal Gaussian distribution, i.e., harmonic or periodic high-frequency components are not represented. If it is necessary to determine whether the vibration signal meets this condition, it is possible to calculate two parameters by which it may be determined whether the distribution is represented by a Gaussian distribution. It is a skewness and kurtosis. For the purposes of vibration diagnostics is mainly

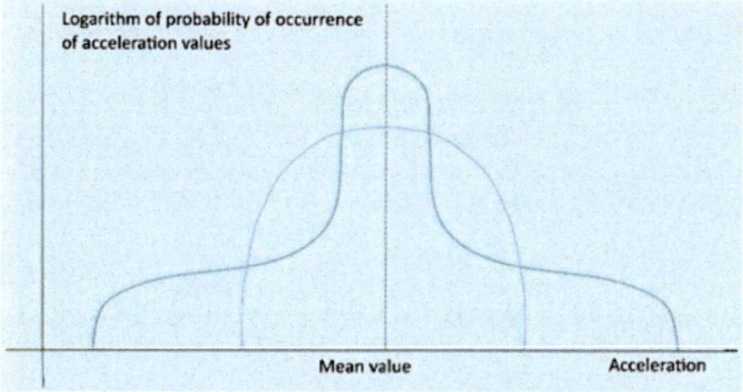

Fig. 3.20 Comparison of signal distribution for good and damaged bearing

Table 3.9 Kurtosis factor value and the corresponding bearing condition

3–4	Good condition
5–8	Early damage
9–13	Serious damage
>13	Potential accident

used kurtosis. Skewness is affected by symmetry of dividing the signal where this characteristic is not so important for the evaluation of the state. In practice, this parameter is called Kurtosis parameter [10] (Fig. 3.20, Table 3.9).

3.2.4 Diagnostics of Roller Bearings—Envelope Analysis

Another common method of vibration analysis is Envelope analysis, which belongs to modulation methods group. The input signal is processed by modulator, which creates signal envelope. Thus, the energy contained in the signal is increased artificially and it is possible to apply for example effective value measurement (ENV RMS), which is more sensitive to damage than the effective value calculated from the signal before modulation. Envelope analysis is a method that not only indicates damage to the bearings, but in association with FFT analysis, it also enables to determine which part of the bearing is damaged. Inner and outer rings, rolling elements, and bearing cage are identified for this purpose. As each of these components has different relative speed with respect to the shaft, it is possible to determine the frequencies where the failure shows. The following equations are used to calculate the failure frequency [10]:

Inner ring:

$$BPFI = RPM\, N_b (1 + \frac{B_d \cos \beta}{P_d})/2 \tag{3.10}$$

Outer ring:

$$BPFO = RPM\ N_b(1 - \frac{B_d \cos \beta}{P_d})/2 \tag{3.11}$$

Ball (roller):

$$BSF = RPM\ P_b(1 - \left(\frac{B_d \cos \beta}{P_d}\right)^2)/2B_d \tag{3.12}$$

Cage:

$$FTF = RPM(1 - \frac{B_d \cos \beta}{P_d})/2 \tag{3.13}$$

where:

Nb number of balls or rollers
Bd ball or roller diameter
Pd bearing pitch diameter
β contact angle degrees

The method is based on measuring the shock impulses generating from the breach of the trajectory along which rolled back the balls or rollers. Envelope analysis lies in the modification of the input signal through a high-frequency filter and envelope detector. This signal is ready for use of FFT analysis and for determination of the fault frequency. The signal is further processed by envelope modulator for ensuring the positioning of repeated shock pulse [11].

When the modulated signal is processed into an FFT spectrum, it appears in a recurring frequency shock pulse, whereas in the modulated shocks not having the character of harmonic signal, it appears generally a series of harmonic signals. It is also common for the failure frequency that it is modulated as a carrier frequency, typically a rotational frequency.

3.2.5 The Correlation of Parameters Measured on Rotary Machine After Reparation of Equipment of the Pulp Production

Based on the cooperation with our workplace practices, we showed the demand for real assessment of the technical condition of equipment of the pulp production (Fig. 3.21) due to previous severe accident on the device (a few week after the repair operation). It was also requested a trends assessment of selected dynamic parameters, e.g., the existence of the risk of damage to the technical system in a short time.

Fig. 3.21 Sample of environment, operating conditions, and ambience of technology workplace

The measurement and assessment of technical device were carried out in accordance with the recommendations of standard STN ISO 10816-3.

The basic parameters of the machine and the description of the operating characteristics are as follows:

- The rotor speed of mixer—874 rpm (14,6 Hz).
- Nominal power Pn = 140 kW.
- Medium: material of the pulp production – 90% water, 10% fiber.
- The abrasive medium is transported at relatively high speed, which entails high attrition of all contact surfaces. Aggressive environment represents a risk factor for corrosion of metal parts.
- For the pulley is used bearings of SKF 22315 with the impeller 22317.

In the context of bearing diagnostics, envelope analysis was applied.

The following table (Table 3.10) shows the calculation of frequencies for the bearings of construction is CC and E. In addition to these types of designs, SKF company also produces design C, CMA, ECC, EC and CY. During the diagnostic system, it was not known that what kind of bearings (manufacturer, structural design) was used in the last repair of the machine.

Table 3.10 Calculation of frequency for above noted types of bearings

Dimensional type of beering	Modi-fication	RPM	BPFO	BPFI	BSF	FTF
SKF 22,315	**CC** **CC**	1 **14.56**	6.1779 **89.95**	8.8221 **128.45**	2.6772 **38.98**	0.4114 **5.99**
SKF 22,315	**E** **E**	1 **14.56**	6.0872 **88.63**	8.9128 **129.77**	2.4945 **36.32**	0.4059 **5.91**
SKF 22,317	**CC** **CC**	1 **14.56**	6.1449 **89.47**	8.8551 **128.93**	2.6140 **38.06**	0.4100 **5.97**
SKF 22,317	**E** **E**	1 **14.56**	6.5652 **95.59**	9.4348 **137.37**	2.6380 **38.41**	0.4107 **5.98**

3.2.5.1 The Correlation of Parameters Measured on Rotary Machine After Reparation of Equipment of the Pulp Production

Recommended levels of warnings—the alarm for rotating machines of a similar type and power is (Table 3.11):

Table 3.12 contains the measured summation values for each methods of measurement, bearing at 22,315 with pulley, rotor bearing:

All measured data show the value of dynamic signal below the recommended limit A1-warning. The measured values refer to the following fact:

- steady state, without any change in the frequency range to 1 kHz and above 10 kHz,
- in frequencies 1–10 kHz a slight increase of about 10%.

Figure 3.22 shows a record of trend graphs—Trend HFD, 3× measuring machine: emergency, after repair (January 2014) and repeated measurement of trend values (April 2014). After the repair, there was a stable condition recorded, without increasing signal.

Table 3.11 Recommended alarm levels

ALARM 1—Warning	4.5 (mm/s g gE)
ALARM 2—Danger	7.1 (mm/s g gE)

Table 3.12 The resulting values for the trend analysis method

Measurement method of dynamic signal	January 2014	April 2014	Evaluation of operating state of the rotor
HFD 40 kHz (gHFD)	0.79	0.89	Good
Acceleration 10 kHz (g)	3.61	3.59	Satisfactory
EnvAcc do 1 kHz (gE)	2.64	2.43	Satisfactory
EnvAcc do 10 kHz (gE)	3.91	4.48	Satisfactory
EnvAcc do 20 kHz (gE)	1.54	1.83	Good

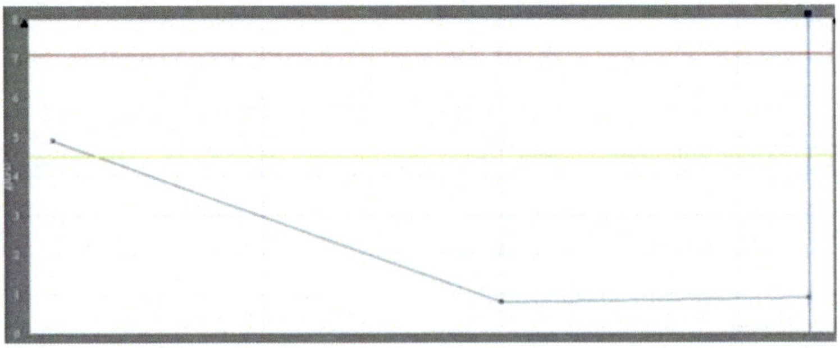

Fig. 3.22 The progress of trend graph HFD

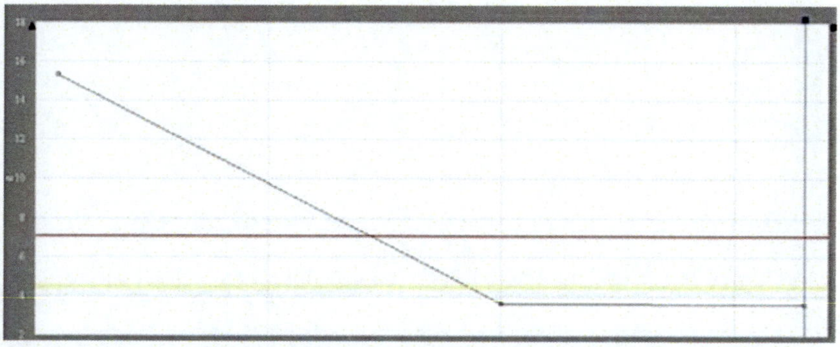

Fig. 3.23 The progress of trend graph Acc

Figure 3.23 shows the record of trend graphs—Acc trend, 3× measuring machine: emergency, after repair (January 2014) and repeated measurement of trend values (April 2014). After the repair, there was a stable condition recorded, without increasing signal.

Figure 3.24 is a graph of the trend—the trend Env Acc to 10 kHz, 3 measuring machine: emergency, after repair (January 2014) and repeated measurement (April 2014). It is visible a slight increase in signal approx. 10%, on the border of recommendations A1—warning.

The records of FFT spectrums for the frequency region to 1 kHz are presented in Figs. 3.25 and 3.26.

In the Fig. 3.25 can be observed low amplitude with the bearing frequency (73 and 146 Hz BSF 6 Hz FTF)—a sign of running storylines.

In April 2014 (Fig. 3.25), there is a slight improvement, e.g., the amplitudes corresponding decrease in the frequency of the bearing.

It may be stated a good deposit condition with no signs of damage and without degradation of the operational status of the contact surfaces.

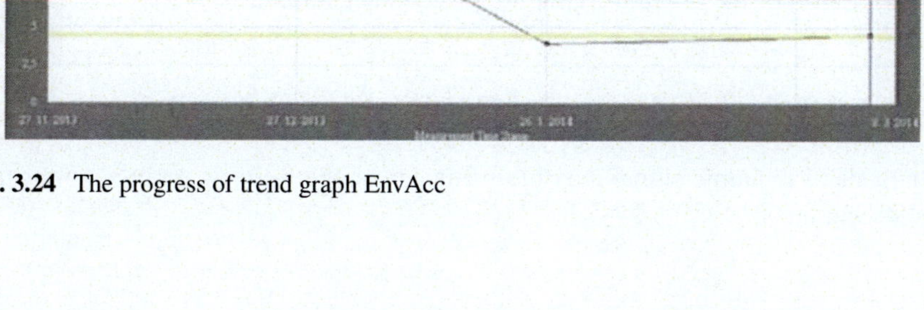

Fig. 3.24 The progress of trend graph EnvAcc

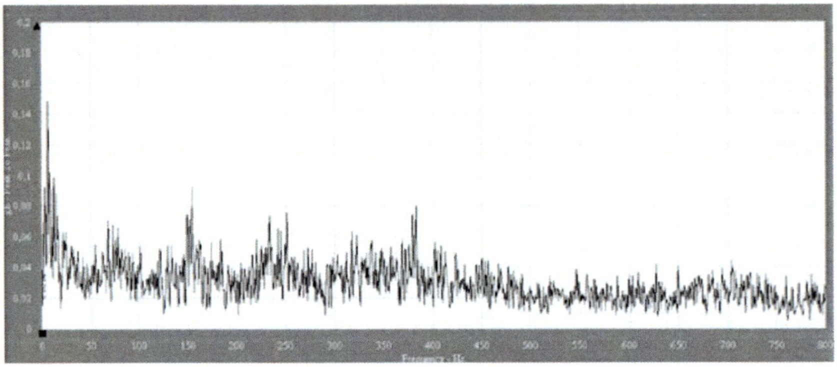

Fig. 3.25 Record FFT spectrum, frequency range up to 1 kHz—January 2014

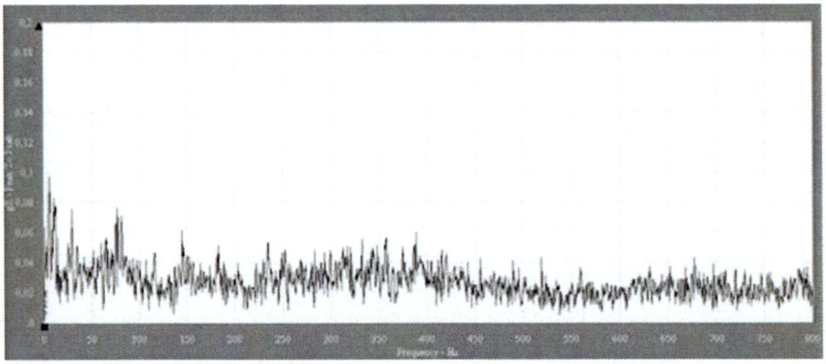

Fig. 3.26 Record FFT spectrum, frequency range up to 1 kHz—April 2014

Data for the FFT spectrum in the frequency region of 10 kHz is presented in Figs. 3.27 and 3.28.

In January 2014 can be observed low amplitude with the bearing frequency (6 Hz FTF)—a sign of running storylines (Fig. 3.27). Summation level in terms of high-frequency "noise" reaches the limit A1-warning.

In April 2014 can be observed low amplitude with the bearing frequency FTF—a sign of running storylines (Fig. 3.28). Summation level in terms of high-frequency "noise" reaches the limit A1-warning.

Good condition of the assembly, no signs of damage, and no significant deterioration of the operating condition of contact surfaces.

3.2.5.2 Analysis of Measurement Results—Discussion

Analysis of dynamic signal measurements across the frequency range points to the following facts:

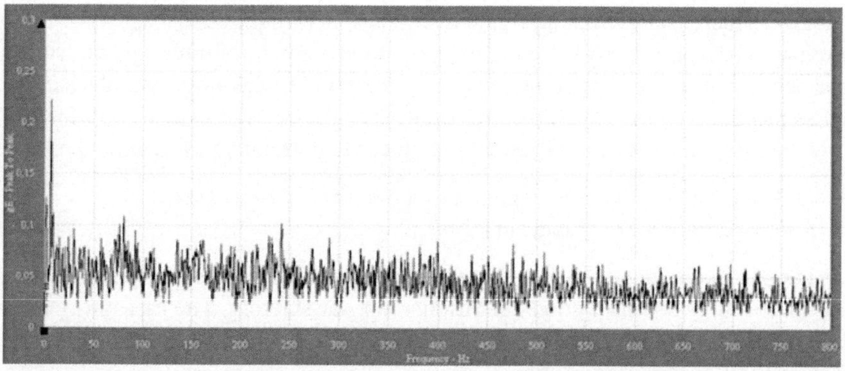

Fig. 3.27 Record FFT spectrum, frequency range up to 10 kHz—January 2014

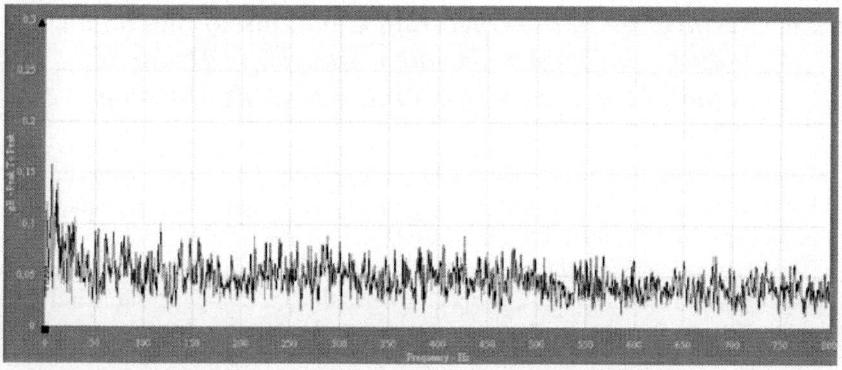

Fig. 3.28 Record FFT spectrum, frequency range up to 10 kHz—April 2014

- good correlation methods used in low frequency up to 1 kHz (velocity, EnvAcc),
- good correlation methods in the frequency range above 10 kHz (EnvAcc, HFD),
- slight discrepancy of 10% signal in the 1–10 kHz, may be caused by non periodic variable signal that is associated with running-engine process (contact surfaces).

Summation values measured the dynamic parameters decreased after repair 4–5 times. The standard rule for basic setup of alarm levels of dynamic signal for rotating machines is:

- signal rise of 200% compared to the reference value (new status, condition after repairs), which means the alarm value (alarm 1—warning),
- In case of further increase in the trend analysis beyond this limit, it is an A2-risk, then, preparatory to repair the measured node.

To verify the measurement method to set binding limits A1 and A2, it would be advisable to carry out repeated trend measurements with an interval of 1 month. In case of an increase in dynamic signal over 200%, it is required to make a correction—exchange of bearings. Then implement the measurement of micro geometry contact surfaces (orbit of bearings), in particular, roughness, waviness, roundness, and the cross profile.

For surface attrition over 1 mm, it is required to correlate the measured parameters:

- FFT and time of dynamic signal, especially in the 1–10 kHz,
- micro-geometry of contact surfaces.

In conclusion, it can be said that the state of the monitored equipment—the pulp blender is after repair in good condition, resp. satisfactory. The measuring methods correlate well with each other; the values after repair are significantly lower and 3 months after repair are stable.

3.2.6 *Verification of the Operating Condition of Stationary Industrial Gearbox Through Analysis of Dynamic Signal, Measured on the Pinion Bearing Housing*

Based on the request of our partner, we have carried out the measurement and analysis of the operating condition of the bearing (analysis of dynamic signal measured on pinion bearing) of the stationary industrial gearbox (Fig. 3.29).

The gearbox is designed for operation in paper industry, in heavy operating mode, especially with respect to rotating speed (thermal balance—generation and removal of heat). The analysis focused on the bearing of the pinion following its dismantling

Fig. 3.29 Condition of gearbox TSA 370-06-1 (two stage, bevel-spur gearbox) after repair

from a failed gearbox. After repair, measurements were performed in order to determine the dynamic properties of the bearings. The values measured on the machine after repair were compared in two operating modes—depending on motor rotation speed.

Measurement and analysis were performed in accordance with the recommendations of standard STN ISO 10816-3 (measurement of absolute vibrations of rotating machines on non-rotating parts).

Figure 3.29 shows the condition of the gearbox following the completed repair. On installation of the pinion (input bevel gearing), the following parameters were set:

- tooth backlash in the range of 0.22–0.23 mm
- axial pre-tension, friction torque of the pair of tapered roller bearings of the pinion 0.5–0.7 Nm
- deformation of the output shaft in radial direction 0.01 mm/1000 N

3.2.6.1 Results of Measurements and Completed Experiments

Basic parameters of the measured component.

- **Gearbox type**: two-stage, bevel-spur
- **Input speed**: max. 2400 rpm
- **Number of pinion teeth**: 20
- **Method of lubrication**: oil circulation through gear pump, feeding oil to each lubrication point (bearings, gearing), PG-based gear oil
- **Pinion bearings**: SKF 32,317 J2 Q (at gearing), SKF 32,314 J2 Q (at coupling)

Input pinion is installed on a single tapered roller bearing. Table 3.13 shows the calculation of failure frequency for outer ring (BPFO), inner ring (BPFI), rolling element (BSF), and bearing cage (FTF).

Analysis of a Failed Bearing After Dismantling

During outage of the paper-making machine, the gearbox was installed and placed in normal operation. The bearing failure occurred in the course of the first 50 h of machine operation. After dismantling from the machine and subsequent analysis of the bearings, damage to pinion bearing SKF 32,314 J2 Q (at the coupling) was found. The following image documents the extent of damage to the bearing.

Figures 3.30, 3.31, 3.32, and 3.33 show visible signs of damage due to excessive thermal loading, signs of tempering of the bearing material, especially on the inner ring at the guide flange (rib). The same condition is found on rolling elements (also at the position of large flange).

Table 3.13 Basic frequency of the gearbox pinion bearing components

Industrial gearbox TSA 370-06-1				
Motor speed: ***1080 rpm*** *Rotation frequency of input shaft (electric motor–coupling–gearbox pinion):* ***18 Hz***				
Bearing type	**BPFO**	**BPFI**	**BSF**	**FTF**
32,317 J2 Q	116.59	171.41	44.41	7.29
32,314 J2 Q	116.53	171.47	44.32	7.29
Tooth frequency: ***360.00 Hz***				
Motor speed: ***1448 rpm*** *Rotation frequency of input shaft (electric motor–coupling–gearbox pinion):* ***24.13 Hz***				
Bearing type	**BPFO**	**BPFI**	**BSF**	**FTF**
32,317 J2 Q	156.31	229.82	59.54	9.77
32,314 J2 Q	156.24	229.89	59.42	9.77
Tooth frequency: ***482.67 Hz***				

Fig. 3.30 Extent of damage to bearing SKF 32,314 J2 Q

Based on the results of the analysis, it is possible to conclude:

- a significant reduction of the hardness of the surfaces supporting axial loading and subsequently—due to transfer of heat, also in axial direction
- flattening of the bearing material
- typical signs of damage in the first hours of machine operation, so-called bluing
- an estimated operating temperature of the bearing components at the point of axial loading during failure, temperature in excess of 350 °C

The bearing cage SKF 32,314 J2 Q shows an intense wear on ribs (Fig. 3.33), loss of material from the cage rib up to fracture, tipping of rolling elements. The cage

Fig. 3.31 Extent of damage to bearing SKF 32,314 J2 Q—inner bearing ring

Fig. 3.32 Extent of damage to bearing SKF 32,314 J2 Q—condition of the rolling elements

Fig. 3.33 Extent of damage to bearing SKF 32,314 J2 Q—condition of the bearing cage

Table 3.14 Measured temperature of the pinion after 6 h of operation in stable mode

	rpm 1080	rpm 1448	Note
Duration of operation in stable mode	6 h	6 h	
Oil supply	In the direction towards the loading flange	In the direction towards the loading flange	
Oil flow through bearing	1 dm^3/min	1 dm^3/min	
Stabilized temperature of the shaft at bearing	73 °C	93 °C	***Recommended limit 110 °C***
Stabilized temperature of the oil pan of the gearbox	60 °C	78 °C	***Recommended limit for PG oil is 96 °C***
Delta T—oil pan gradient/pinion shaft	13 °C	15 °C	

was loaded by excessive force for which the bearing is not designed under normal conditions.

Diagnostic Measurements, Analysis of Dynamic Signal

After the gearbox repair, diagnostic measurements were performed with emphasis on stabilized temperature of the pinion and analysis of the dynamic signal from the input pinion. The measurements were performed for two operating modes of the gearbox:

1. mode—input speed 1080 rpm
2. mode—input speed 1448 rpm

Table 3.14 shows the measured temperature of the pinion after 6 h of operation in stable mode Since the start, we observed increasing temperatures and stabilization on certain components of the machine (oil temperature on bearing intake, temperature of oil pan, temperature of gearbox casing above oil level, gearbox case temperature at the pinion bearing 32,314 J2 Q, shaft temperature at the inner ring of the pinion bearing 32314J2Q. The measurements were performed for different values of oil flow (the table shows data for bearing oil flow 1 dm^3/min) as well as for various methods of oil supply to the bearing (supply from small flange, supply from thrust flange).

Rotary machines generate vibrations during their operation. Analysis of vibration signal enables us to acquire information on the technical condition or operating conditions of the monitored machine. With respect to overall measured vibration values, it was necessary to identify the recommended limits with respect to the requirements of technical standard STN ISO 10816-3(6). For different operating modes of the machine (considering speed, loading, performance), it is possible to set the warning limits (alarms) also on the basis of general recommendations (signal increase by 200% compared to reference value).

Default alarm levels:

ALARM 1—Warning	4.5 (mm/s g gE)
ALARM 2—Danger	7.1 (mm/s g gE)

When measuring the velocity of vibration according to the recommendation of ISO 2372 to assess mechanical vibrations in low-frequency range, the following types of failures can be identified: dynamic imbalance, misalignment of shafts at couplings, bent shaft, low rigidity of the foundations or mechanical release—backlash (play), structural resonance of frames, boxes, and foundations [12, 13, 14].

When measuring the vibration acceleration envelope (EnvAcc) in high-frequency range, the following failures can be identified: fatigue damage to the bearings, damage to contact surfaces of the bearing with "metallic" contact, severe operating conditions of the bearings (e.g., due to excessive mechanical vibrations), inadequate friction—poor lubrication in some cases (generally the same applies to gearing) [12, 13, 14].

The following diagnostic instruments and devices were used for the applied diagnostic measurement of vibrations (data collection) and analysis:

- frequency analyzer and Data Collector MicrologGx,
- SW environment Aptitude Analyst,
- Instrument manufacturer SKF Condition Monitoring (USA). For details on technical parameters of the instrument, see www.skfcm.com
- vibration sensor, accelerometer Wilcoxon Research, Model SKF786M, sensitivity 100 mV/g, frequency range 1–20.000 Hz,
- information on vibration sensors at www.wilcoxon.com and www.skfcm.com.

The following methods were applied:

For assessment of vibrations in low-frequency range:

- measurement method, velocity (vibration velocity); unit, mm/s, and RMS detection,
- pursuant to recommendation of ISO10816-3, FFT spectrum for frequency range 10–800 Hz,
- measurement method, enveloping acceleration, unit, gE, PtP detection, for frequency range 50–1000 Hz, FFT spectrum, and time recording.

For assessment of vibrations in high-frequency range:

- measurement method, acceleration (vibration acceleration), g, PtP detection, for frequency range up to 16 kHz, FFT spectrum, and time recording.
- measurement method Enveloping Acceleration, unit gE, PtP detection, for frequency range 10–20 kHz, FFT spectrum, and time recording.

The following table (Table 3.15) shows the increase in the total summary values of measured parameters for increased input speed.

Table 3.15 Increase in the total summary values of measured parameters for increased input speed

Measured parameter	Operating mode rpm 1080	Operating mode rpm 1448	Relative increase by (%)
Velocity (mm/s) up to 1,000 Hz	0.81	1.14	41
EnvAcc (gE) up to 1,000 Hz	2.28	2.87	26
Acceleration (g) up to 16,000 Hz	1.47	2.68	82
EnvAcc (gE) up to 10,000 Hz	2.10	4.19	99
EnvAcc (gE) up to 20,000 Hz	0.66	1.75	165
HFD (gHF) up to 40,000 Hz	0.35	0.97	177

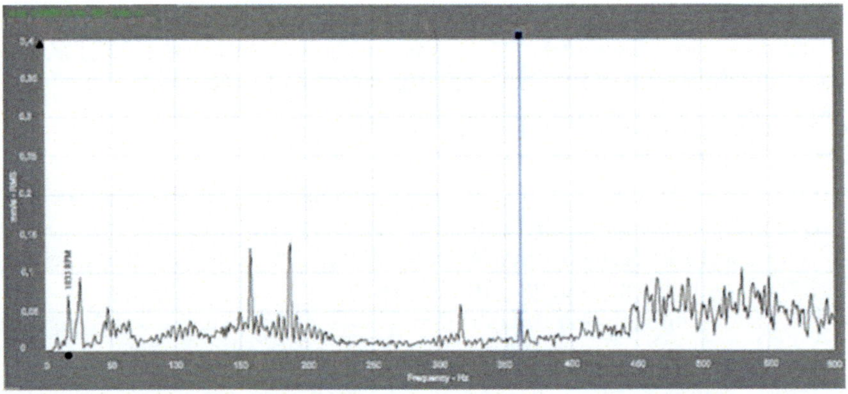

Fig. 3.34 Velocity (mm/s), measuring position 1 V—1. mode rpm 1080

The subsequent steps of analysis were oriented towards comparison of dynamic signal, measured on gearbox (measuring position 1 V—pinion bearing at gearing, measuring point 1H—pinion bearing at coupling) during performed tests (Figs. 3.34, 3.35, 3.36, and 3.37).

In case of measurement of vibration velocity in Figs. 3.34 and 3.35, in both cases of applied modes, dominant amplitudes are:

- RPM input
- engagement of the gearing of the 1st and 2nd stages of the gearbox.

Amplitude and overall signal level are under the recommended limit of Alarm 1—warning. The increase in the monitored values in the second regime is about 40%.

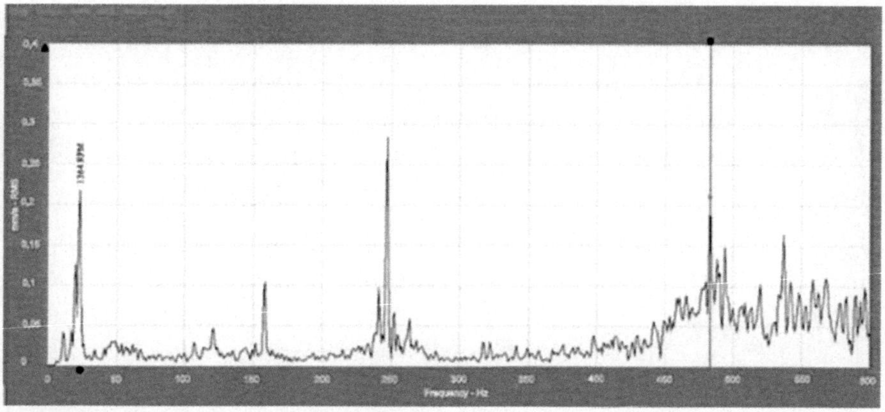

Fig. 3.35 Velocity (mm/s), measuring position 1 V—2. mode rpm 1448

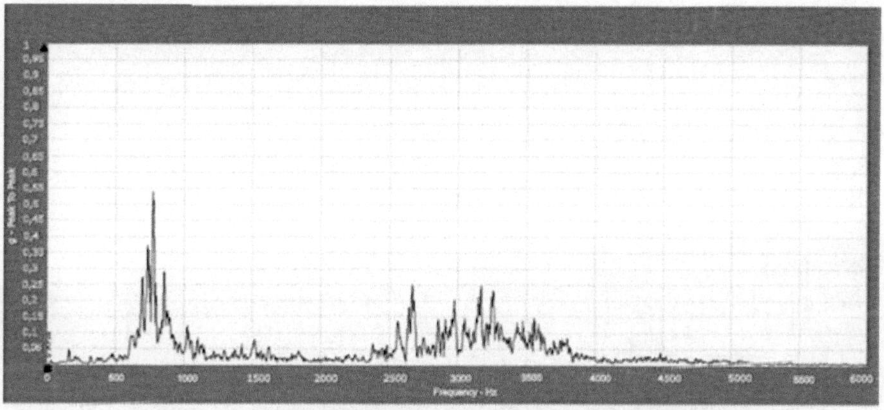

Fig. 3.36 Acceleration (mm/s), measuring position 1 H—1. mode rpm 1080

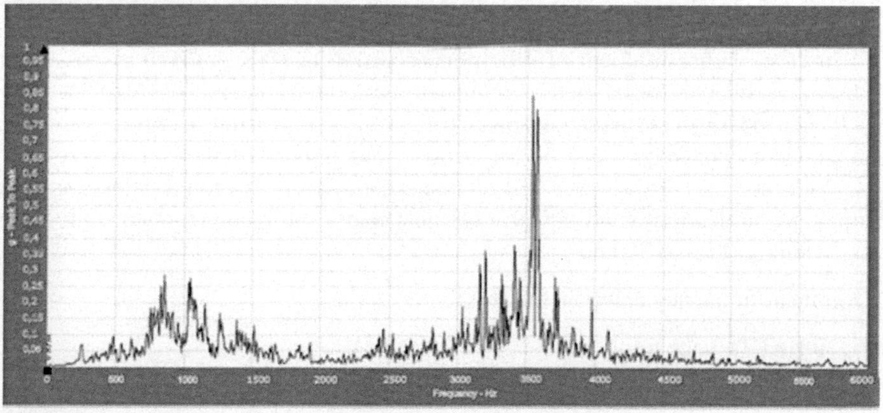

Fig. 3.37 Acceleration (mm/s), measuring position 1 H—2. mode rpm 1448

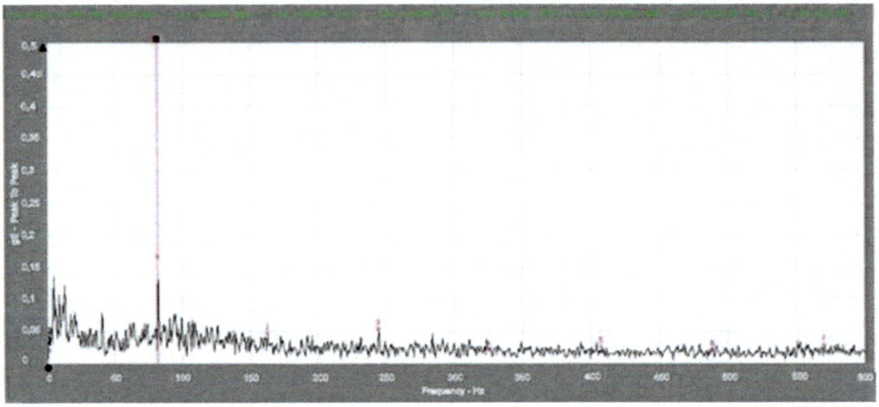

Fig. 3.38 EnvAcc (mm/s), measuring position 1 H—1. mode rpm 1080

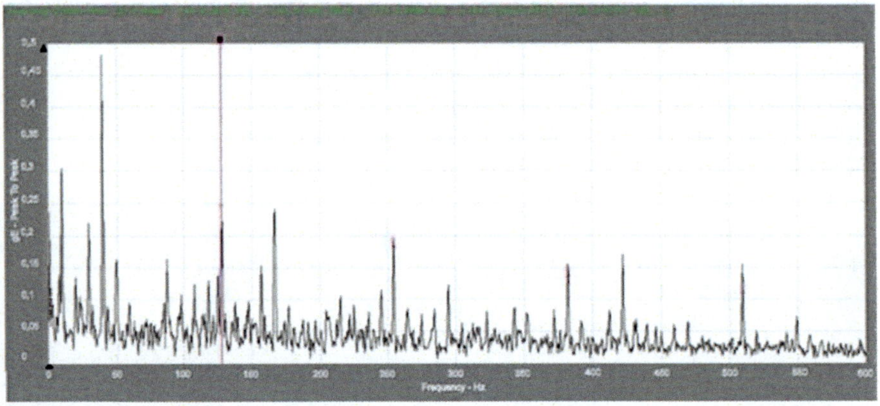

Fig. 3.39 EnvAcc (mm/s), measuring position 1 H—2. mode rpm 1448

In case of measurement of vibration acceleration (Fig. 3.36), it is possible to observe resonance "peak" in the vicinity of 2× tooth frequency 770 Hz.

When measuring in the second mode (Fig. 3.37), it is possible to observe resonance in the vicinity of frequency 2.5 kHz—area of bearing.

With the observed values of envelope acceleration (vibration acceleration envelope), we notice increased signal level, just under the recommended limit Alarm 1—warning (Figs. 3.38 and 3.39).

Dominant frequency components are:

- 10 Hz FTF cage frequency
- 128 Hz frequency of rolling elements 2xBSF and multiples

The measured values represent the indication of adverse operating conditions in bearing 32314J2Q.

3.2.6.2 Results of Analyses and Discussion

Analysis of the bearing following its dismantling after gearbox failure showed the following adverse conditions and signs:

- so-called bluing of the bearing in the first hours of machine operation,
- discrepancy between heat generation and heat transfer away from the pinion bearing,
- excessive increase in the bearing temperature above 350 °C, resulting in metallographic changes of the properties of components (tempering).

In the subsequent step, we performed diagnostic analyses after repair of the gearbox, with stabilized gearbox operation. The results of these analyses can be summarized in the following overview:

- high bearing temperatures at 32314J2Q, close to the recommended limit
- high oil temperatures, close to the recommended limit
- high temperature gradient between bearing and oil pan
- high levels of dynamic signal, especially in frequency range over 2 kHz
- FFT analysis showed a problem in the operation of the bearing, especially in segments of rolling elements/cage

The measured values of dynamic parameters and temperature (after repair) correlate with the condition discovered in the failed bearing (after dismantling from failed gearbox). Overheating of the bearing above the allowed temperature may result in reduction of the original hardness of the material—steel, used to manufacture the bearings. This results in reduction of bearing quality and its failure. This is demonstrated by changed color of the rolling elements, rings and cages. The issue may be addressed by application of lubricant of required viscosity.

In order to determine the limit modes of the operation in the particular application, the following is recommended:

- perform the measurement of the gearbox on paper-making machine, in various operating modes
- determine the so-called reference signal levels, observe the increase in the selected parameters
- perform a trend analysis, including the analysis of gearbox oil (content of iron wear particles in the oil, content of solids in the oil, content of water and corrosive particles, thermal-oxidation degradation and TAN code, content of elements—loss of additives)

The presented study was focused on the application of the analysis of dynamic signal recorded on bearing housing of the pinion in a two-stage bevel-spur gearbox operating in paper-making plant. This operating environment is characteristic by difficult operating conditions, especially with respect to speed (thermal balance—generation and transfer of heat). The purpose of the application of the envelope analysis was to verify the operating conditions that may cause damage to the technical condition of the observed object and under critical circumstances, it may result in failure—involving great financial losses for the partner company. The gearbox was repaired due to a previous breakdown. Pinion bearing failure occurred in the course of the first 50 h of machine operation. The results of the used analyses correlated with the condition of the dismantled bearing determined during repair of the gearbox. This results in recommendations for the operator—leading to adjustment of the operating parameters of the gearbox.

One of the requirements to achieve long service life and reliability of the gearbox is the reliable operation of the bearing. Among the most important operating parameters that require monitoring and that must be kept with in recommended limits are [12, 13, 15, 16]:

- assembly procedures, setting of radial play during installation, setting of pre-tension in the clamp joint between conical bushing/pin of the roller, prescribed quality of geometry and micro-geometry of the connecting elements (see the recommendations of bearing manufacturers).
- friction and condition of the lubricant, especially with respect to solids (quantity, size, type), iron wear particles (intense wear increases radial play of the bearing and changes the micro-geometry of the contact surfaces). Very important in this case is also the content of water and corrosive particles (for limits, see recommendations of the bearing manufacturers and lubricant manufacturers—for example maximum iron content 100 ppm, max. water content 200 ppm…).
- deviation from stable operation of the bearings in the recommended mode can be monitored by measurement and analysis of dynamic signal (e.g., acoustic emissions, high-frequency absolute vibrations, etc.). It is necessary to repeat the measurements (trend measurements—comparing the changes, e.g., increase in signal by 200%) and also monitor the FFT spectra (expected frequencies of bearings—assignment to the measured signal)
- all observed parameters must be put in correlation with the condition of the contact surfaces of the bearings, subsequently determining the limits for individual methods.

The applied tools of technical diagnostics confirmed the behavior of the observed segment of the technical system under analysis in the particular industrial setting. The purpose of the diagnostics was not just the assessment of the current condition of the gearbox pinion. Very important is also the indication of the condition and organization of the maintenance process in the partner company. It is clearly recommended to focus on proactive maintenance—i.e., continuous monitoring of technical equipment and subsequent control of the causes of failure.

3.2.7 The Dynamic Parameters Correlation Assessment of the Textile Machine High-Speed Bearings in Changed Technological Conditions

The utilization of technological equipment in the textile industry has its particularities mainly due to the presence of chemical and biological factors. In this field, the production process consists of several phases: processing of natural raw materials (cotton, flax, hemp, or sisal hemp), fiber treatment (cleaning, dyeing, bleaching, maceration, suppression, plating, and impregnation), weaving on knitting machines and the final product manufacturing—sewing, knitting. The dust produced in the textile production often contains the residue of chemically treated substances, which is why the dust can be toxic. The development of bearings for spinning machines is closely related to the solutions for the mounting of the most important textile machine elements. Their use is ultimately not limited to textile machines; they can also be utilized in other machinery. Special ball bearings for textile machines are designed for high rotational speeds and relatively low loads. They are characterized by high dimensional accuracy and reliable running, which guarantees their great utility value. These bearings are often supplemented with some other components, e.g., flexible mountings. In some cases, they make complex integrated bearing units that enable technologically and economically more efficient production of textile machines.

Bearings are often confronted with adverse conditions during operation, which can lead to premature damage.

3.2.7.1 The Description of the Research Problem

Based on the partner company's request, we have undertaken the diagnostics of high-speed bearings of the spinning machine applicable in textile production (Fig. 3.40a). The machine manufacturer and technology buyers demand high operational reliability and high bearing durability.

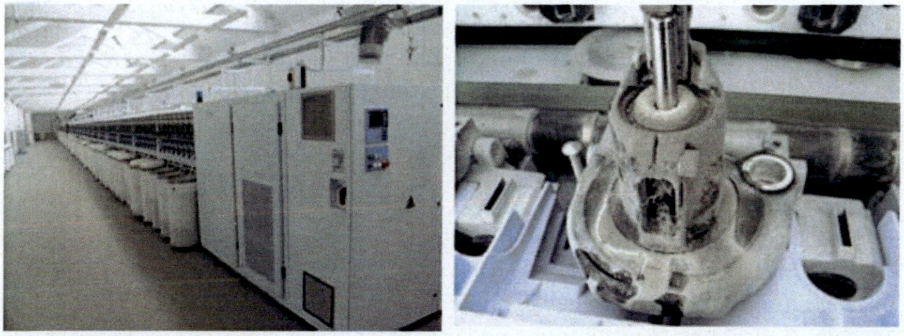

Fig. 3.40 Spinning machine applicable in textile production, **a** The environment of the spinning machine, **b** The pollution of a spinning machine spindle by textile production dust

A spindle of textile machinery, especially of spinning machines, is a shaft, at one end of which is a fixed spinning rotor and the other end is connected to the drive. The shaft is mounted in a special double-row ball bearing, the outer sleeve of which is supported by elastic elements fitted into a carrier body that is attached to the machine frame. At a higher weight of the spinning rotor, the shaft acts as a flexible body, large vibrations are generated, and extensive dynamic forces affect the bearings in that they reduce their lifetime.

In this context, the spinning spindle high-speed bearing has undergone diagnostic measurements and assessment during its being in operation. Measurements—dynamic data collection—were performed in two modes of spinning spindle operation: in real machine operation, with roller bearing speed of 120,000 rpm, and through comparing selected spindles at the test station in laboratory conditions at our workplace at the speed of 90,000 rpm.

3.2.7.2 The Description of the Research Problem

The measurement and analysis of vibrations has an unmistakable place in the field of diagnostics of rotary machines. The machine generates vibrations when in operation. Such a vibration signal mediates information that can point to [17, 18]:

- Machine operating conditions, in particular power and dynamic loads, magnetic phenomena, lubrication mode, and the like,
- Damage to the machine node or component, including the possibility of assessing the level of its severity and its effect on the next operation.

The measurement and spectral evaluation of absolute speed of vibration provides information on basic types of rotational machine defects. The vibrations detected on the non-rotating parts of the machine (bearing housing, cap, etc.) close to the speed of the machine are analyzed. By evaluating the measurement of various machines and devices, a number of failures can be assessed, but generally this method is particularly suitable for diagnostics in the following cases [17, 18]:

- dynamic imbalance of rotating discs or cylinders,
- non-axial connection of shafts on couplings,
- resonance phenomena and machine operation in the area of critical speed,
- mechanical release and broken machine foundations,
- bent shaft,
- damage to rolling bearings and tooth gears.

In this method, the absolute velocity of the measured surface is evaluated; yet, it is necessary to remember that the product of velocity and mass corresponds to the momentum of the body in mechanics, which is the energy quantity [17, 18].

Measurement and spectral evaluation of high-frequency vibrations up to 20 kHz, especially acceleration of vibration and modern ways of enveloping, are best used in the assessment of the operating state or in the detection of the damage of:

- mountings with roller bearings, especially for speeds above 200 rpm,
- toothed gears as well as multi-stage and planet gearboxes,
- in many cases, lubrication, metal contact identification, or low oil film capacity.

The most suitable method for tracking and evaluating vibration signals is the FFT—fast Fourier transformation method. In mathematical terms, this means that the signal is spread over certain amplitudes corresponding to the different frequency components. In the event that something is wrong with the machine, FFT spectra are able to provide information that helps to locate a fault, to determine its cause, and by means of trending to determine the time period at the end of which the machine will be in a critical condition. Since we know that a certain failure occurs at certain frequencies, FFT spectra are analyzed by observing amplitude changes in these frequency ranges [2, 17].

In order to determine the exact condition of the bearings, a special so-called envelope vibration acceleration technology has been developed. The action of bearings and the engagement of gear wheels, which are of a repetitive nature, produce vibration signals with a much lower amplitude and higher frequencies than vibrational signals generated by the speed or structure. Rotational vibration signals and amplified repetitive signal components are filtered out from defects in bearings. The purpose of the enveloping is to filter the low speed vibration signals associated with the speed and to highlight the signals, to make them separate from the bearing damage that appears in the frequency domain indicative of bearing failures. The reason for the use of envelope technologies is based on the knowledge that rolling away the damaged bearing element causes blows that trigger increased vibrations to the blow frequency but mainly to the resonant frequencies. The most common cases of the envelope application are rolling element bearing defects and gear engagement analysis, where the low amplitude of the repeating vibration signal can be hidden in the vibration noise of the machine originating from the speed and design. For example, if there is a defect in the roller bearing on its outer track, each rolling element passing through this position produces a small repeating signal with a frequency corresponding to the defect in the bearing. However, this vibration signal is so low that it is totally "lost" when measuring total (overall) vibrations, caused by other rotational and structural phenomena. In order to focus on these recurring components in the frequency range typical of bearing defects (e.g., repeating vibrational bearing signals), low-frequency signals are filtered out by means of enveloping, and thus, periodic shock signals stand out and focus on recurring events in the frequency range, corresponding to the defects in the bearing [2, 17].

The envelope analysis is a method that not only indicates bearing damage but, in conjunction with FFT analysis, makes it possible to determine which part of the bearing is damaged. For these purposes, the outer and inner rings, rolling elements, and a bearing cage are distinguished. Since each component has a different relative speed with regard to the shaft, it is possible to determine the frequency at which the failures are manifested. The method is based on the measuring of shock pulses that arise when the trajectories are disturbed by balls or rollers. The envelope analysis is based on the adjustment of the input signal using a high-frequency filter and an envelope detector. In this way, the signal is ready for applying FFT analysis and for determining the fault-indicating frequency. The signal is further processed by the envelope modulator, which ensures the highlighting of repetitive shock pulses. When the modulated signal is processed into the FFT spectrum, repetitive impulse frequencies will appear therein. Since modulated shocks do not have the character of a harmonic signal, a number of harmonic components will also appear. It is also common for fault-indicating frequencies to be modulated on carrier frequencies, typically rotational.

3.2.7.3 The Characteristics of Measurements, Equipment, and Methods Applied

In the diagnostics of the spinning machine spindle high-speed bearings, absolute vibrations were measured according to the recommendation of STN ISO 10816-3 norm.

For the diagnostic measurement of vibrations (data collection) and subsequent analysis, the following were used:

- Vibration sensor, Wilcoxon Research accelerometer; model, SKF786M, sensitivity 100 mV/g, and frequency range 1–20.000 Hz,
- Frequency Analyzer and MicrologGX Data Collector (SKF Condition Monitoring, USA),
- SKF @ptitude Analyst software environment

Measurement methods:

To assess vibrations in the low frequency range, the following were used:

- Velocity (speed tune); unit mm/s, and RMS detection,
- In accordance with ISO10816-3, the FFT spectrum for the 10–800 Hz frequency range,
- Enveloping acceleration, Eg; PtP detection for the frequency domain 50–1000 Hz; and FFT spectrum and time recording.

To assess the vibration in the high-frequency area, the following were applied:

- Measurement, acceleration; unit, g; PtP detection; frequency range up to 16 kHz; FFT spectrum, and time recording,
- Enveloping acceleration, Eg; PtP detection; frequency range up to 10 kHz and up to 20 kHz; FFT spectrum and time recording.

Methods used for assessing acoustic emission and ultrasound are:

- Spectral Energy Emission See (SKF), See unit, PtP detection, frequency range up to 600 kHz, FFT spectrum, and time Recording,
- High-frequency detection (HFD), the analysis of vibration (acceleration) in the frequency range of 40 kHz

These methods can generally be implemented to assess the lubrication, bearing capacity of the oil film, metallic contact, seizing or abrasion of the contact surfaces (bearings and cog system).

3.2.7.4 Total (Summarized) Values of Measured Quantities

Measurements were made at the odd positions of the spinning machine rotors labelled L1 to L415.

Figures 3.41 and 3.42 show the measured vibration values for the EnvAcc measurement method in the frequency band of 0.5–10 kHz. The red color indicates positions that display high-frequency signal values above the recommended Alarm 2 limit (A2–Danger). The yellow color marks elevated values above the recommended Alarm 1 limit (A1—warning).

The values of alarm levels (primarily for noise limits, sound pressure, acoustic emission and high-frequency acceleration) are determined by the following ways:

- by agreement between a supplier (manufacturer) and customers,
- when the signal increase over 200% compared to the reference (agreed) level measured on the normal product quality and design,
- by comparing the dynamic characteristics of competing manufacturers–suppliers, respectively, by comparing the measured values with the normal level of the same products in the local market.

In the specific case of defined alarm, set points were based on an analysis of signal levels of 95% of the measured values of controlled amount of the same types of mechanisms.

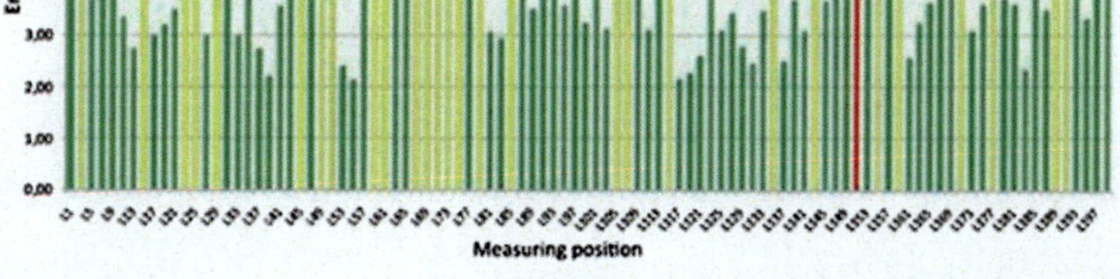

Fig. 3.41 Measured vibration signal values for the EnvAcc method—envelope of acceleration up to 10 kHz at odd positions L1–L199—February 2016

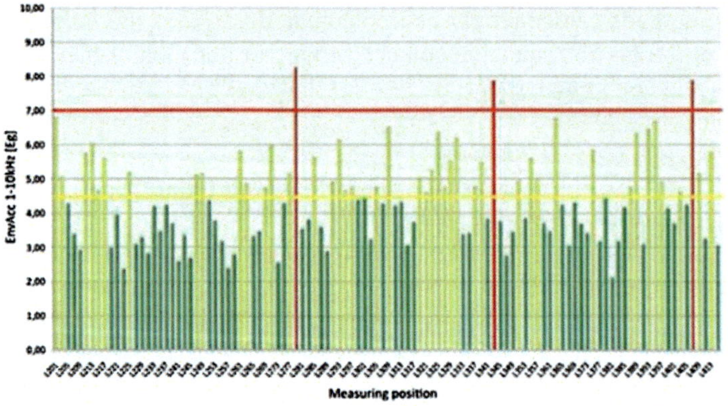

Fig. 3.42 Measured vibration signal values for EnvAcc method—Envelope of acceleration up to 10 kHz at odd positions L201–L415—February 2016

3.2.7.5 Comparison of Measured Dynamic Signal Before and After Speed Increasing

The graph in Fig. 3.43 shows the statistics—the average value of the vibration signal (EnvAcc 1–10 kHz) for changed operating conditions of the machine (in 2015, spindle speed 105,000/min; in 2016, increase in spindle speed to 120,000/min). It is obvious that a 15% increase in speed has been reflected by a 50% increase in dynamic signal (from 2.79 to 4.24 after 1500 h of operation)—Fig. 3.43.

This trend chart shows an increase in the summation value of the signal in the 1–10 kHz bandwidth. The values correlate with the deteriorated state of microgeometry (wear) of the spinning rotor bearing paths (roughness, corrugation, etc.). The values

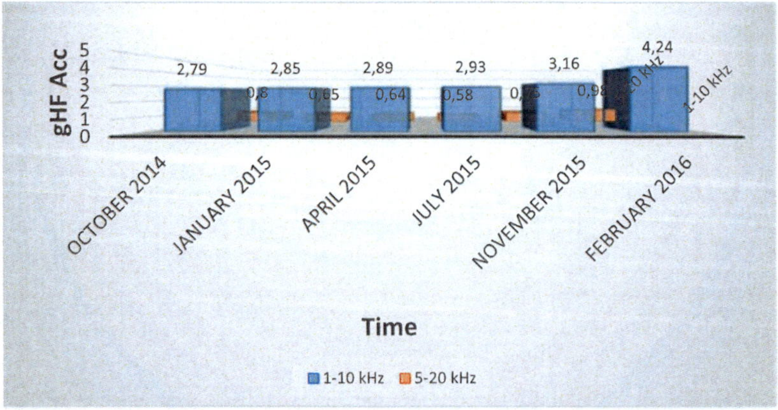

Fig. 3.43 Average vibration signal values (EnvAcc 1–10 kHz) after speed increasing

have increased from 2.8 to 4.2 (gE). The method of 5–20 kHz bandwidth measurement is intended for monitoring the friction mode (lubrication state, bearing capacity of oil film, stalling, metal contact, mechanical impurities and abrasion particles in lubricant, etc.). Signal increase was significantly lower than in the case of the first method. It can be assumed from the above that the dominant part of the measured signal (and the transmitted information content) lies in the 1–5 kHz bandwidth.

3.2.7.6 The Impact of Spindle Cleaning and Bearing Turning on Dynamic Parameter Values

By visual inspection of the spindles, intense clogging of the space between the rotor bowl and the bearing face has been detected. Impurities cause high vibration and rotor braking, which is accompanied by a decrease in speed. Both factors cause significant technological problems during fiber spinning. The following measured data indicate the condition before and after cleaning the rotors. Figure 3.44 shows the measured values of the vibration signal for the EnvAcc method 1–10 kHz—the 1st measurement before cleaning the rotors and the 2nd measurement after their cleaning.

The influence of cleaning and rotation of the bearings on the speed is given in Table 3.16.

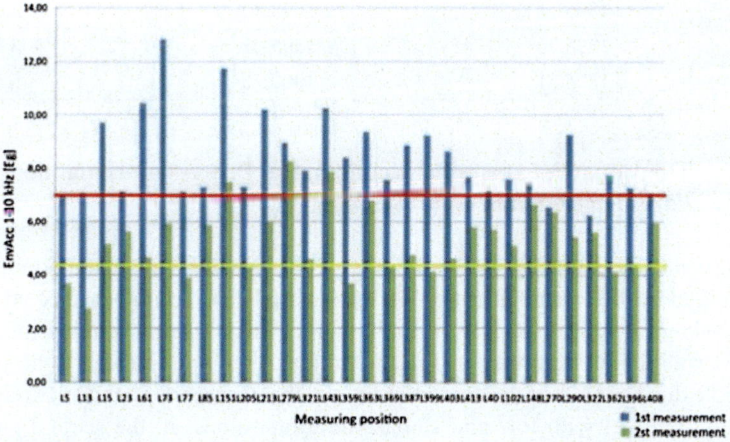

Fig. 3.44 Measured values of vibration signal for EnvAcc—envelope of acceleration up to 10 kHz before and after rotor cleaning and bearing rotation

Table 3.16 Impact of spindle cleaning and rotation of bearings on the speed

Measuring site	Speed before cleaning [Hz]	Speed after cleaning [Hz]	Difference
L5	1883.08	1886.31	3.24
L13	1886.50	1885.59	**−0.91**
L15	1886.74	1886.11	**−0.64**
L23	1884.95	1888.42	3.47
L61	1883.32	1885.99	2.67
L73	1886.25	1889.23	2.98
L77	1887.07	1889.42	2.35
L85	1888.70	1888.22	**−0.48**
L151	1891.54	1891.69	0.15
L205	1888.52	1889.67	1.16
L213	1892.44	1894.41	1.97
L279	1895.14	1895.41	0.27
L321	1895.57	1895.06	**−0.51**
L343	1897.36	1898.00	0.64
L359	1892.42	1892.77	0.36
L363	1896.22	1898.24	2.02
L369	1895.23	1898.17	2.95
L387	1897.98	1899.01	1.03
L399	1896.11	1898.80	2.69
L403	1898.36	1895.07	**−3.30**
L413	1895.61	1900.70	5.10

3.2.7.7 The Impact of Unstable Technological Process on Dynamic Parameter Change

The influence of an unstable technological process on dynamic parameter change is exemplified by the measurement of high-frequency vibrations on a selected spinning machine rotor—measuring station L290, before cleaning and rotating the bearings and the rotor. The recording in Fig. 3.45 shows the amplitudes of the measured vibration during 33 s of rotor operation. In the first part of the recording, rotor operation is quiet with low and stable vibration values. In the second part of the recording, the signal is higher, and the rotor running is unstable (variable amplitudes of the signal).

Figure 3.46 shows the recording of the vibration amplitudes—1 s from the first—quiet area of the operation of the monitored rotor. The same recording from the second—unstable area of operation of the observed rotor L290 is shown in Fig. 3.47.

The FFT spectrum from the first—quiet part of the recording (Fig. 3.48) points to the dominant rotation frequency of the T125 bearing rotor. The amplitude at the

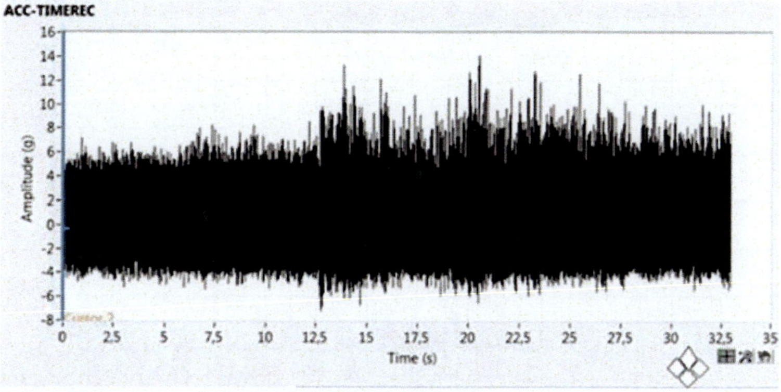

Fig. 3.45 Recording of measured vibration amplitude for L290

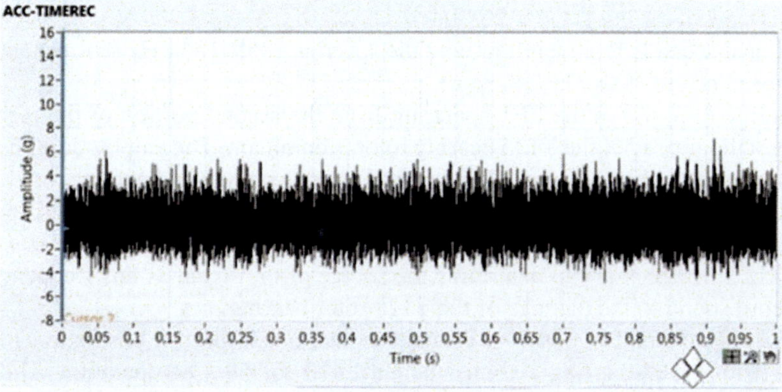

Fig. 3.46 Vibration recording—1 s from the quiet L290 rotation recording area

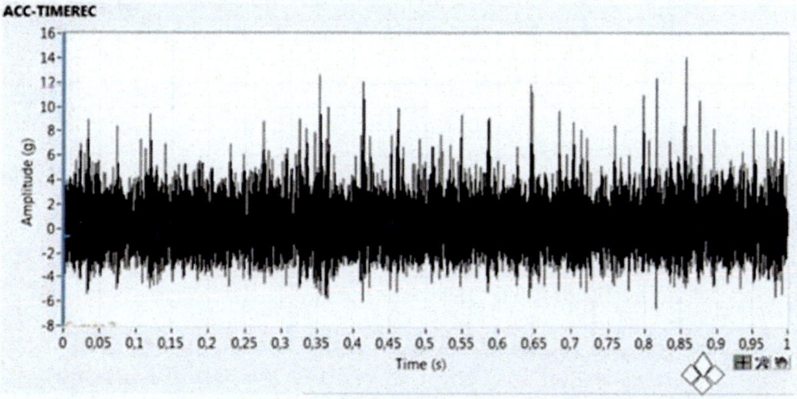

Fig. 3.47 Vibration recording—1 s from an unstable L290 rotor recording area

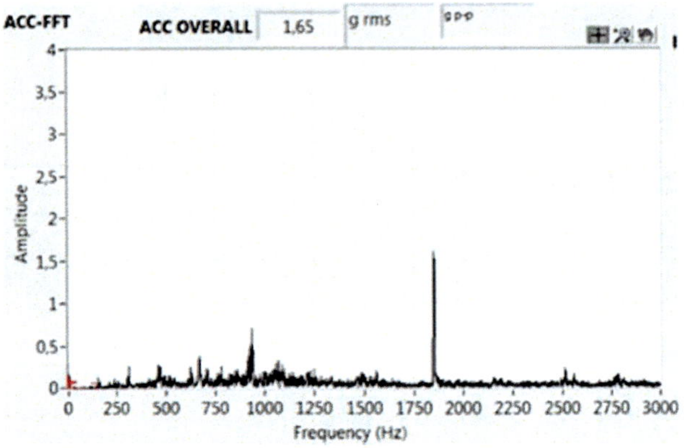

Fig. 3.48 FFT spectrum from the first—quiet part of the recording

rotational speed is relatively low, and the speed is 1850–1854 Hz (corresponding to the speed of 111,000–111,240 rpm).

Figure 3.49 shows the FFT spectrum from the second portion of the recording. The rotation speed of the T125 bearing rotor is dominant. The amplitude at the speed frequency is by 130% higher. The speed frequency is reduced and has the value of 1850–1854 Hz (111,000–111,240 rpm), which calls for rotor braking. In the spectrum is also a second vibration source with the frequency of 312 Hz, 2 × 312 Hz, and 3 × 312 Hz. It is necessary to determine the source of the signal as this frequency does not correspond to the damage of the T125 rotor bearing.

Another example of observed changes in vibration is the case of spinning machine rotor with the measuring station labelled L413 for 60 s of operation. The graph

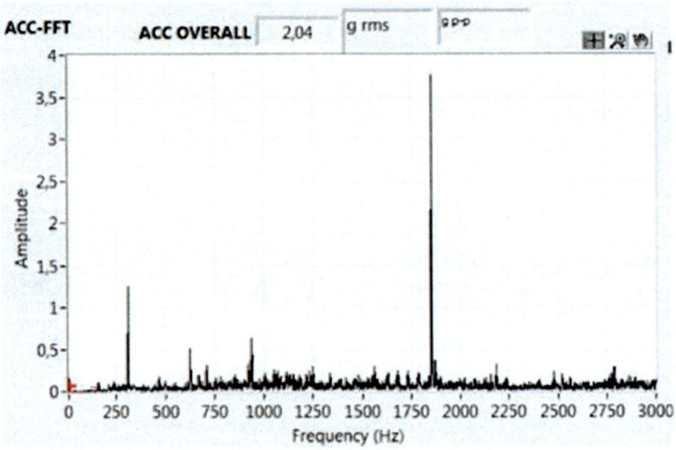

Fig. 3.49 FFT spectrum from the second—unstable field of operation of the monitored rotor L290

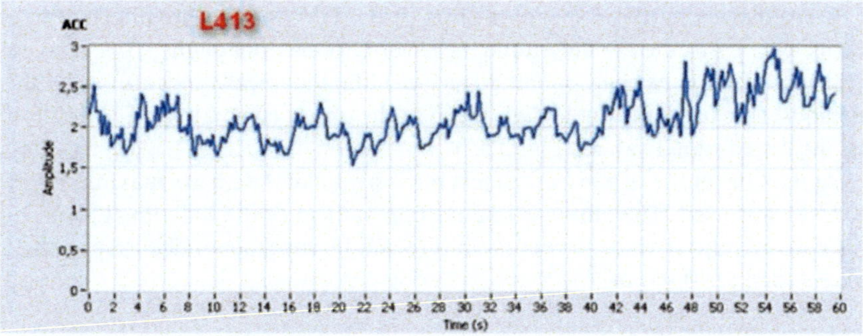

Fig. 3.50 HF vibration sum value—L413

Table 3.17 Calculated expected bearing frequencies T125

Rotor speed–bearings T125/L413		Calculated expected frequencies [Hz]			
[Hz]	[RPM]	FTF (cage)	BSF (rolling body)	BPFO (outer ring)	BPFI (inner ring)
1	60	0.3599	1.6402	2.1594	3.8406
2000	120,000	719.8	3280.4	4318.8	7681.2
1917	115,000	689.8	3143.7	4138.9	7361.2
1833	110,000	659.8	3007.0	3958.9	7041.1

depicted in Fig. 3.50 shows the measured signal—the sum value of HF vibrations. The T125 bearing rotor shows signs of relatively high dynamic "instability" and variable signal. The total value varies from 1.5 gHF to 3.0 gHF, which is a symptom of an unstable technological process. The dominant frequency sources are identical to the previous measurement example at station L290.

The following table (Table 3.17) contains the calculation of the expected T125 bearing rotor frequencies—L413.

3.2.7.8 The Diagnostics of Selected Spinning Rotors at a Test Station Under Laboratory Conditions

Based on the detection of a high dynamic signal in real operation, our partner was recommended to dismantle rotors with high HF vibrations. After removing the rotors from the working positions on the machine, the following operations were carried out:

- cleaning the space between the spinning bowl and the bearing face (contaminated by clogged fiber)
- dynamic balancing of rotor and spinning bowl

Consequently, at the test station (Fig. 3.51), HF vibration monitoring measurements were performed to determine the severity of bearing damage (wear of contact surfaces—rolling bearing running surface). In the measurements, only the HF methods were used due to the high speed of the observed bearings. The graph shows the measured signal values (Fig. 3.52).

At the time of the experiment, a test measuring stand with the maximum speed of 90 kPM was used. The aim was relative comparison of the individual rotors—their absolute vibrations value. Eventual failure and its subsequent occurrence shall also be recorded at 90 k rpm.

Fig. 3.51 Test station for measuring high-speed bearings

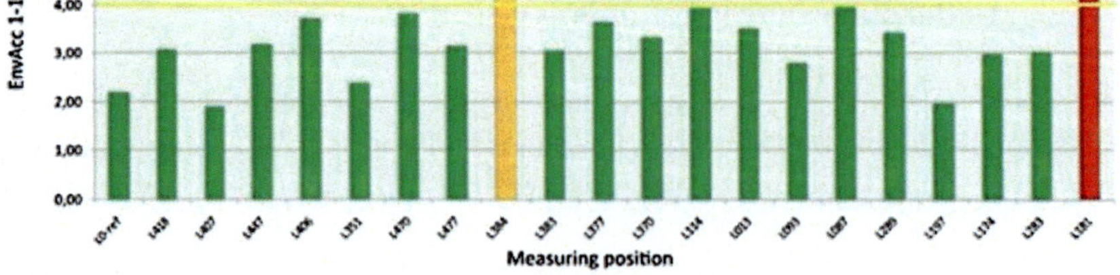

Fig. 3.52 Measured vibration signal values for the EnvAcc method—envelope of acceleration up to 10 kHz performed at the test station at 90,000 rpm—February 2016

The HF signal analysis revealed damage to the rotor bearing only in two cases (see Fig. 3.52). Other bearings were sent back to the machine operator. We proposed that they should be re-installed and used without restriction, in the proper mode.

At the same time, we recommended that the following should be visually observed:

- fiber rupture (frequency, a sign of speed deceleration, or unstable rotor speed);
- temperature, vibration, noise, and the like.

3.2.7.9 The Evaluation of Results of the Applied Measurements and Analyses—Discussion

Based on the measurements and analyses of the dynamic signal on the textile machine, it can be stated:

- The increase in machine speed (rotor speed) by 15% resulted in a statistical increase in measured dynamic parameters by 50%,
- The increase was caused by two main sources: the signal from the rotor bearing and the signal from the technological spin "instability"
- At the same time, new drawbacks have occurred: rotor clogging by fiber during combing and subsequent braking and speed deceleration, a signal from a damaged drive belt, variable dynamic imbalance of the rotor, and at the same time an uneven bowl wear caused by the fiber friction
- All of the above mentioned operating failures occurred after 1500 h from transition to the new technology.

In order to achieve the desired stability of the machine, it is necessary to implement a measure to achieve the improvement and eliminate the identified drawbacks. Otherwise, the described deficiencies will result in a significant reduction in the durability of the rotor bearings. We can expect that due to damaged bearings the number of spindle breaks will be approximately 10 times faster and 10 times more frequent than in regular operation.

The study presented the case of application of the technical diagnostics tools in verifying the influence of changes in the textile spinning process on the level of service life of the spinning rotor bearings. The measurement showed a statistical increase in the measured dynamic parameters by 50% when the spinning machine spindle rotor speed was increased by 15%. In particular, the measured vibration values showed that technological instability (rotor clogging by fiber during combing and subsequent braking and speed deceleration, a signal from a damaged drive belt, variable dynamic imbalance of the rotor, and at the same time an uneven bowl wear caused by the fiber friction) had a noticeable impact. On the basis of this, we have provided our partner with suggestions on how to achieve the desired life of the high-speed spindle rotor bearings of the observed spinning machine.

3.3 Combination the Diagnostic Methods as Suitable Tool for Increasing an Effectivity of Determination the State of Mechanical Nodes

An important parameter of any machinery is its flawless operation. In order to ensure the trouble-free operation of machine is need to realize how important non-destructive and online diagnostics these machines. In this case, the gearbox was diagnosed and the generator of small hydroelectric power station, which were flooded during the extraordinary floods. So this study will be to deal with diagnosis of rotating machinery. Measurements of vibration and ultrasound have been transferred in the following periods:

16/03/2010—the routine planned measurement (operation is normal)
02/07/2010—measurement after the floods
14/07/2010—the repeated measurement after the floods (12 days of operation)
20/07/2010—measurement after repair (exchange the bearings)
17/09/2010—the repeated trend's measurement

It is important to mention the gearbox was dismantled, cleaned, and re-lubricated immediately after the floods. Repeated measurements of vibrations, however, pointed to a hidden defect, namely in one measuring point (roller bearing NU 322). There was a performed the diagnostics of defective bearing too (tribotechnical diagnostic of lubricant, measurement of corrugatedness, measurement of roundness, measurement of roughness of its functional surfaces), after the exchange the bearings.

3.3.1 The trend's Measurement of Vibrations

A. **The trend's measurement of vibrations in frequency band to 10 kHz**

The measuring of absolute vibrations was conducted on non-rotary parts of the gearbox. The following graphs show the change of total values of vibration, depending on the time.

The results of a trend's measurement (presented in Fig. 3.53) refer on increased of vibrations so the value of vibrations exceeded the alarm1 during the defective period. The value of alarm1 is 4,5 gE and presents a warning. These values were measured in the horizontal direction only. The results of measurement of vibration in the vertical direction are presented on Fig. 3.54 and shows on even higher values of vibrations. Their value exceeded alarm 2 during the defective period; it means over the 11 gE. This alarm presents a danger.

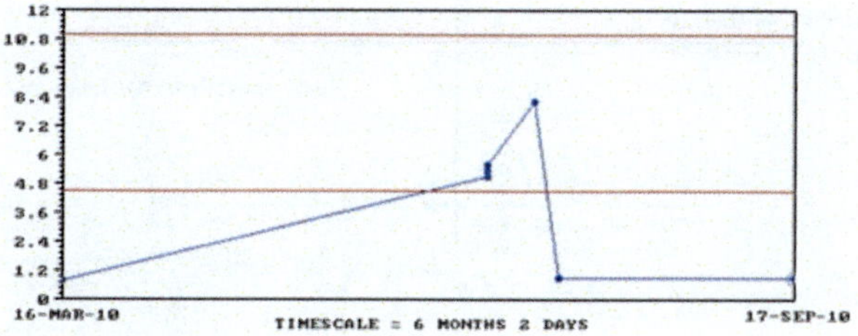

Fig. 3.53 The trend´s measurement of enveloping acceleration to 10 kHz, gE, PtP

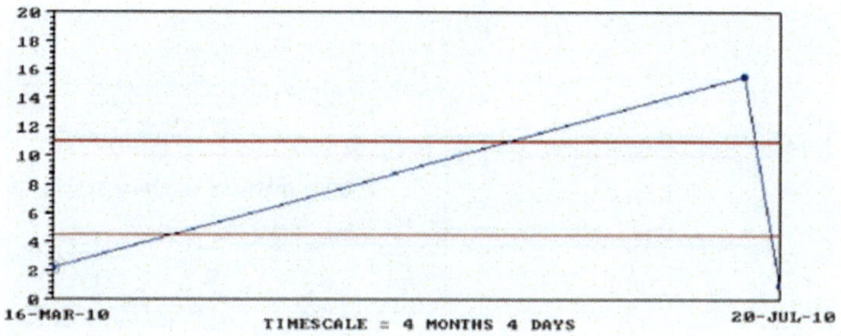

Fig. 3.54 Same measuring in vertical direction

B. Determination the technical state of bearings in a frequency band around 40 kHz

The measuring of vibrations in this frequency band refers to the defective state also. The values of vibrations were close to the alarm 2 during the most critical period, which shows the Fig. 3.55.

C. FFT spectrum in frequency band to 10 kHz

The following graphs to show the measured values of vibrations (FFT spectrums) according to a measured period. Dependence of the measured values and the trend's

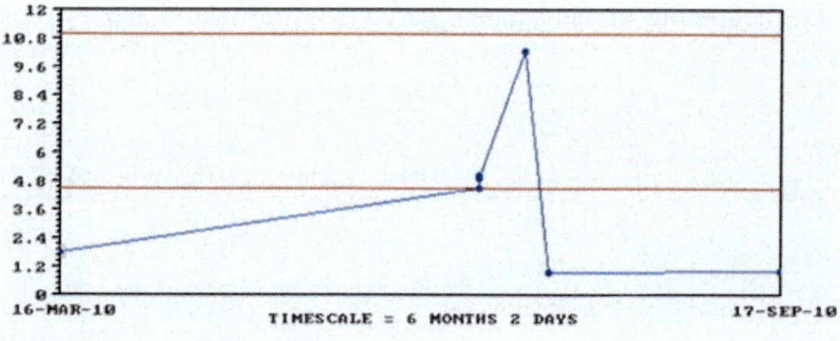

Fig. 3.55 Trend HFD versus time trend

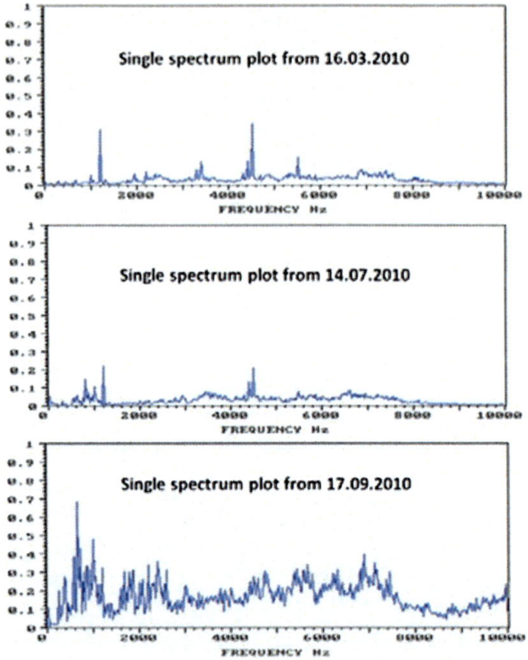

Fig. 3.56 FFT spectrum in frequency band to 10 kHz

measurement is visible. The amplitude of oscillation is demonstrably higher in a critical period than in the period prior to flooding (Fig. 3.56).

D. **FFT spectrum in frequency band to 40 kHz**

Each manufacturer of rolling bearings delivers the information sheet to his products, which also contains the frequency values.

These values to provide with information about particular defect of bearing and they are in Table 3.18.

Since the range of these values is moving around a low hertz values there was a necessary to make the measurement of vibration in the range up to 400 Hz. The resulting values of these measurements are shown in the following graphs. Measurement at the beginning of the trend's period does not show any extreme vibration (Fig. 3.57).

Table 3.18 Frequency values for roller bearing NU 322

Particular parts of bearing	The value of frequency [Hz]
Outer ring	66.7
Inner ring	96.6
Bearing roller	33.2
Bearing cage	5.1

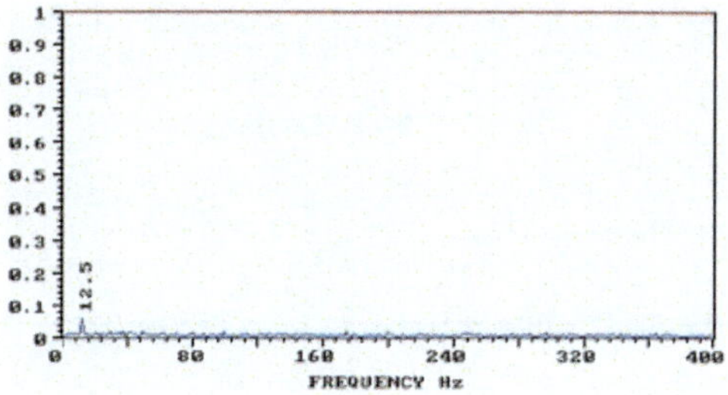

Fig. 3.57 FFT spectrum in frequency band to 40 kHz 16.03.2010

The results of subsequent measurements, taken during the first half of July (Fig. 3.58), clearly show at the fluctuations of amplitude in 66.5, 96.5, and 67 Hz. It refers to the defects of bearings rollers, inner ring, and outer ring.

Graphical results from another measurement show an evident reduction of vibration after the removal of defect in the later period (Fig. 3.59).

E. **The trend of ultrasonic emission in a frequency band around the 300 kHz**

Results of the trend's measurement of ultrasonic emission clearly show on a defective state during the critical period as you can see in Fig. 3.60, when the value of ultrasonic emission to high exceeds a limit value. But this also shows at an improvement of technical condition after the repair.

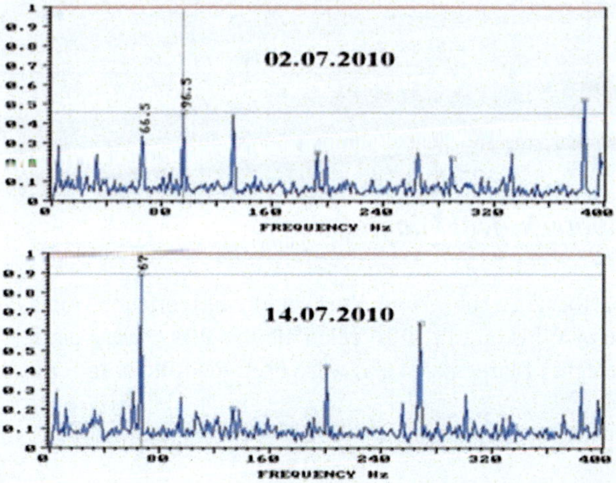

Fig. 3.58 FFT spectrum in frequency band to 40 kHz 02.07.2010 and 14.07.2010

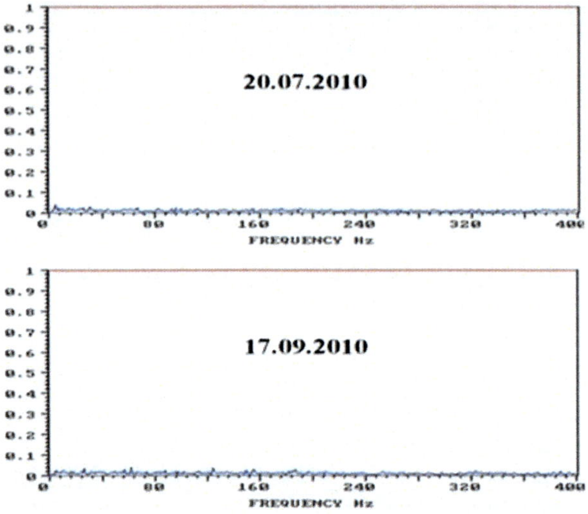

Fig. 3.59 FFT spectrum in frequency band to 40 kHz 20.07.2010 and 17.09.2010

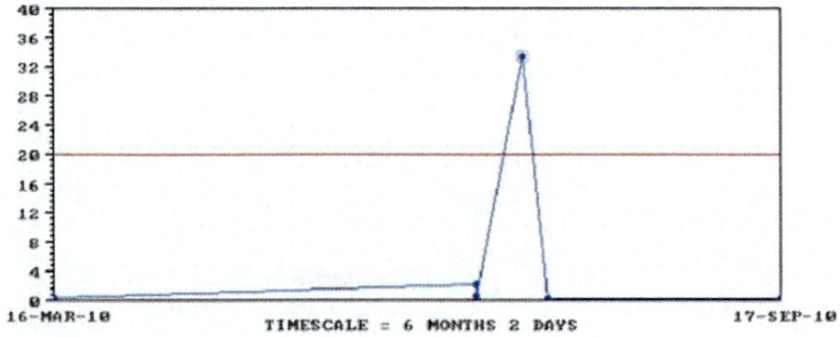

Fig. 3.60 The trend's measurement of ultrasonic emission

3.3.2 *Tribotechnical Diagnostic*

There was performed a subscription the sample of lubricant from a defective bearing to determine of WPC and PLP. To calculate the WPC (wear particle concentration) and PLP (percent of large particles) were used the following formulas:

$$WPC = \frac{D_L + D_S}{1800} \times 100\% \tag{3.13}$$

$$WPC = \frac{D_L + D_S}{1800} \times 100\% \tag{3.14}$$

Where the DL is a number of large particles, DS is a number of small particles and the value of k is derived from Table 3.19.

Table 3.19 The value of variable k

D_L	<99	<199	<299	<399	<499	<599	<699	<799	<899
k	0.01	0.1	0.2	0.3	0.4	0.5	0.6	0.7	0.8

Table 3.20 The measured values and calculated results

	D_L	D_S	WPC	PLP
Bearing NU322	375	17	22%	28%

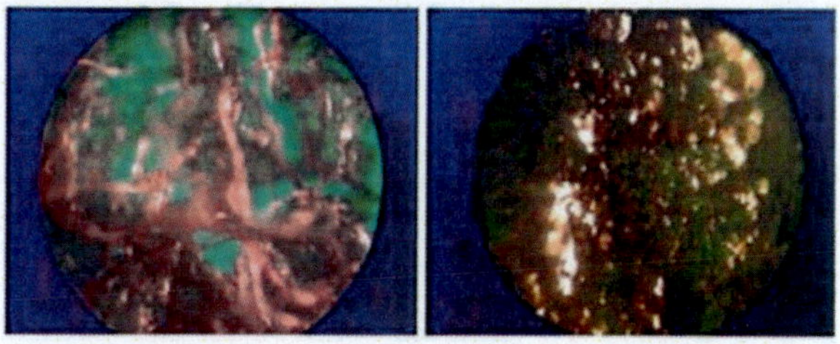

Fig. 3.61 Samples of wear particles

Following table shows the measured values as their calculated results. According to these results can be said that the level of wearing (WPC) is good but the intensity of wearing (PLP) is perilously high (Table 3.20).

Consequently, a pherographical analysis was performed and then the photos of wear particles were created (Fig. 3.61).

The picture on the left side (Fig. 3.61) shows an excessive number of oxidative particles Fe_2O_3, which have an abrasive character. The picture on the right side shows on the abrasive particles. Their size is under the 250 μm, and their occurrence is inflicted by the presence of hard particles. The hard particles make the scorch on the surfaces and they are an index of catastrophic wear.

3.3.3 *The Surface Analysis*

The defective bearing was disassembled and cleaned after the tribotechnical analysis. There were performed the measurements of corrugatedness, the measurement of roundness, and the measurement of roughness subsequently.

The measurement was performed on the functional surfaces of defective bearing (circumferential area of bearing roller, external area of inner ring, and inner area of outer ring), which are shown in Fig. 3.62.

Fig. 3.62 Functional surfaces of defective bearing

3.3.4 *The Measurement of Roughness*

The measurement of roughness was performed on the visible damaged surfaces and visible undamaged surfaces. The device of Mitutoyo SJ-301 was used for measurement of roughness. The Mitutoyo is portable device directed to a practice with an integral printer (Fig. 3.63 and Table 3.21).

Measured parameters:

– the center-line average (CLA) roughness (Ra),

$$Ra = \frac{1}{l}\int_0^l |Z(x)|dx \tag{3.15}$$

Table 3.21 Measured values of roughness

Bearing roller	Damaged surface	Ra = 3.42
		Rz = 19
	Undamaged surface	Ra = 0.11
		Rz = 0.95
Inner ring	Damaged surface	Ra = 5.17
		Rz = 30.34
	Undamaged surface	Ra = 0.4
		Rz = 2.72
Outer ring	Damaged surface	Ra = 1.97
		Rz = 12.71
	Undamaged surface	Ra = 0.44
		Rz = 2.17

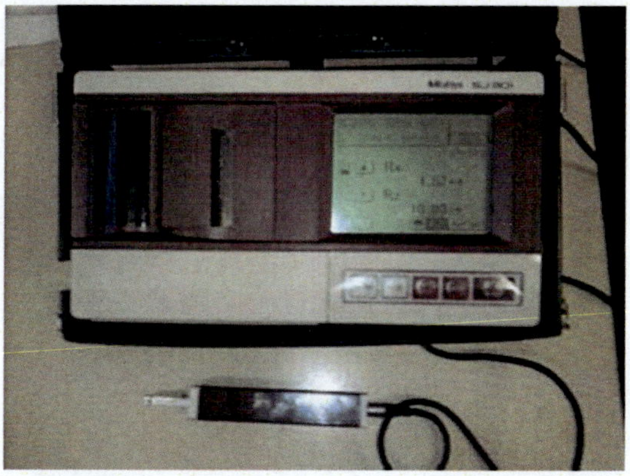

Fig. 3.63 Mitutoyo SJ—301

– vertical distance (Rz) between the highest peak (Rp) and the lowest (Rv) valley of the unfiltered profile

3.3.5 The Measurement of Roundness

The device Mitutoyo Ra—120 was used for the measuring of roundness and is shown at Fig. 3.64. This device serves on measurement of deviation's shape (Figs. 3.65 and 3.66).

The results of the measurement of roundness performed on a bearing roller shows at the angularity. The results of the measurement of roundness performed on a bearing roller shows at the angularity and it considering to a small number of waves. The number of waves is in the range from three to five.

The outer ring has higher value of roundness and it has an inconstant course. This ring has an undesirable trigon, which did the strokes in a bearing. This method clearly confirms the critical state similar as the methods before.

Fig. 3.64 Mitutoyo RA—120

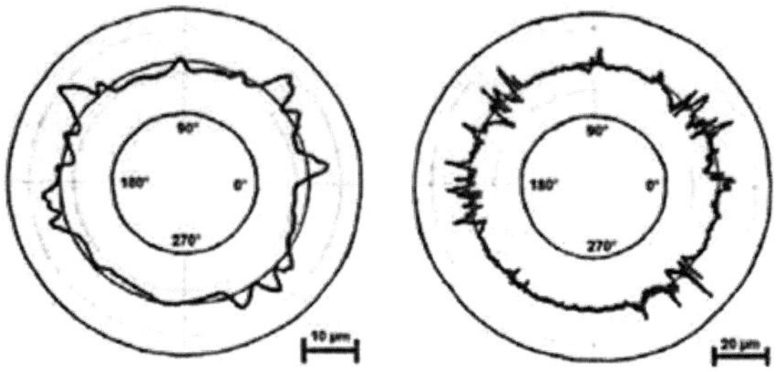

Fig. 3.65 Roundness: ΔR = 8.63 μm Corrugatedness: f = 17.3 μm

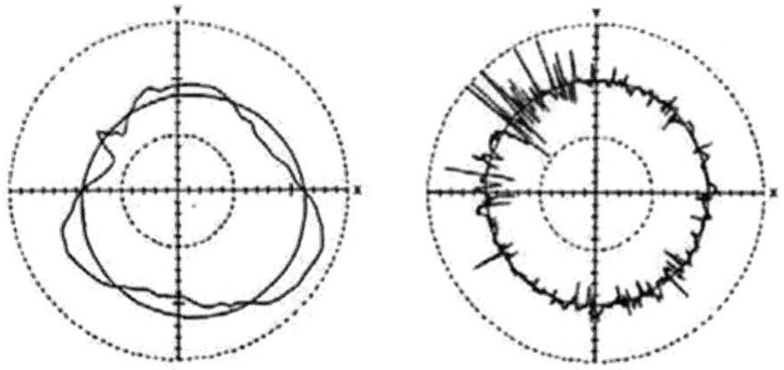

Fig. 3.66 The roundness: P + V = 10 μm, the corrugatedness: P + V = 24.5 μm

3.3.6 Results of Analyses and Discussion

An online watching of technical state of machines, which are nonstop working, is very important part of their current maintenance. The study shows at their reliability and accuracy in the determination of accurate place of defect. other diagnostic methods suggest it too. The contribution further shows at the direct dependence between mentioned diagnostic methods, where the all three nondestructive methods (vibration, ultrasonic, and tribotechnical) agreed on the place of defect.

Using all the diagnostic methods is not very effective especially from an economic aspect. Therefore, it is necessary to choose the most effective method, and then, if the selected method is not sufficient, it is necessary to choose the other one.

3.4 Application of Methods of Technical Diagnostics by Assessment of Oil Filling Condition in the Process of Running-In of Planetary Gearbox

On the basis of cooperation applied between our work location and the practice, there arised the demand of implementation the analysis of the oil condition in the planetary gearbox of machining center type VB-715A. Specifically, the ZF is a two-stage planetary gearbox providing optimal torque at low and high spindle speed range.

The aim of the diagnostics of the system under review was:

- to identify the condition of the abrasive particles in oil (quantity, type, size)
- to assess the condition of the contact surfaces in the running-in process (analysis of wear particles)
- to assess the condition of the lubricant
- correlation, quantification of measured parameters—characteristics of the recommended limits

3.4.1 Materials and Methodology

Implemented analysis of gear oils has been preferentially oriented to identify the wear particles produced during running-in period and in the first hours of operation of contact surfaces. At the same time, the correlation between total content of Fe (iron) particles in oil and accrual of kinematic will in tooth meshing of toothed ring of the gearbox was monitored. In quantification of the content of ferrous wear particles, the impact of the running-in processes on the contact surfaces (bearings, toothing), which can cause negative side effects, was assessed. Side effects may occur in the form of:

- changes of microgeometry of surfaces, wear, respectively abrasive (adhesive) wear,
- contamination of the oil filling by solid particles of abrasive character which cause the wear rate increase and acceleration of the claw will and the will in the bearings of the gearbox.

To assess the vibration in the low frequency area, there have been applied:

- Velocity measurement method (vibration velocity); unit, mm/s; and RMS detection
- as recommended by ISO 10816-3:2009, FFT (fast Fourier transform) spectrum for frequency area from 10 to 800 Hz
- Measurement method, enveloping acceleration (method of measuring the vibration acceleration in the high-frequency area from 10 to 20 kHz through which especially the condition of rolling bearings and gearings is assessed), gE unit, PtP detection, for frequency area from 50 to 1000 Hz, FFT spectrum, and time record

To assess the vibration in the high-frequency area, there have been applied:

- Measurement method, acceleration (vibration acceleration), g unit, PtP detection, frequency area up to 16 kHz, FFT spectrum, and time record
- Measurement method, enveloping acceleration, gE unit, PtP detection, frequency area from 10 kHz and to 20 kHz, FFT spectrum, and time record

Diagnostic methods for the assessment in field of acoustic emission and ultrasound:

- Measurement method, See (Spectral Emitted Energy)—acoustic emissions (SKF), See unit, PtP detection, frequency area up to 600 kHz, FFT spectrum, and time record
- Measurement method HFD—high-frequency detection method (SKF), vibration analysis (acceleration) in frequencies of 40 kHz
- Measuring method is used to assess the lubrication, resistance of the oil film, metallic contact, sticking, and abrasion of contact surfaces (bearings and gearing)

Analyses of lubricating oil:

- Fe wear particles (size, number, type) arising from the use of tooth gears and bearings have been assessed (monitored parameter WPC—wear particles confrontation, PLP—strength of wear and ppm Fe)
- the thermal–oxidative degradation of oil and the contact surfaces is assessed, the particles of carbon type, carbon black, oil oxidation, and the oxidative particles Fe (CCT IR parameter)

Advanced analyzes of lubricating oil:

- sample of lubricating oil Madit X-220 after elapsing 867 h in the test
- monitored parameters of gear oil are listed in text below

In the applied analysis were during testing implemented:

1. repeated (trend) diagnostic measurements of vibration and acoustic emission
2. assessment of the operational condition of bearings and gearings
3. assessment of the friction mode, analysis of wear particles type Fe, and analysis of degradation of gear oils

In Tables 3.22 and 3.23, there are characterized conditions under which the measurements were made.

Table 3.22 Conditions of applied measurements—part A

Number of hours in operation of gearbox [h]	10	109	247	508	950
Output speed [rpm]	7.23 8.96	–	7.59 8.82	7.63 8.62	7.63 8.62
Vibration measurement	Yes	–	Yes	Yes	–
Oil sample for analysis	Yes	Yes	Yes	Yes	Yes
Oil fill Number of hours in operation [h] Oil fill rank	PP-90 10 First	PP-90 109 First	X-220 138 Second	X-220 400 Second	X-220 841 Second

Table 3.23 Conditions of applied measurements—part B

Number of hours in operation of gearbox [h]	30	300	500	755	1005
Output speed [rpm]	5.0	–	5.0	–	5.36
Vibration measurement	Yes	–	Yes	–	Yes
Oil sample for analysis	Yes	Yes	Yes	Yes	Yes
Oil fill Number of hours in operation [h] Oil fill rank	X-220 30 First	X-220 300 First	X-220 500 First	X-220 755 First	X-220 1005 First

3.4.2 Results of the Measurements and Experiments

3.4.2.1 Course of Parameters Monitored—Summing Values—Part A

The parameters of measurements and tests for this part of the analysis are shown in Table 3.22. Tables 3.24 and 3.25 contain comparisons—trends of measured summing (total) values of vibration. It is a measuring point 1 H, V—of tooth meshing on the first stage at a speed of 7 and 9 rpm.

After the initial running-in of surfaces, there has been an improvement—monitored parameters decreased. All monitored parameters of high-frequency vibrations show good, respectively satisfactory, operating condition of tooth meshing of gearing, and bearing conditions.

Tables 3.26 and 3.27 provide a comparison and trends of measured summing values of vibrations at the measuring point 2 H, V—of tooth meshing on the second stage at a speed of 7 and 9 rpm.

Even in the second stage, after the initial running-in of surfaces, the improvement, i.e., decrease in monitored parameters, is shown. All monitored parameters of high-frequency vibrations indicate good operating condition of tooth meshing of gearing and bearing conditions.

Table 3.24 Vibration values measured at the measuring point 1H, VS (1st stage)

Output speed [rpm]	7.23	7.59	7.63	Valuation
HFD [g]	2.4	2.2	2.5	Good
ACC [g]	2.9	2.3	2.7	Good
ENV up to 10 kHz [gE]	3.1	2.6	2.5	Good

Table 3.25 Vibration values measured at the measuring point 1H, V (1st stage)

Output speed [rpm]	8.96	8.82	8.62	Valuation
HFD [g]	3.1	2.6	2.7	Satisfactory
ACC [g]	3.3	2.9	3.2	Satisfactory
ENV up to 10 kHz [gE]	3.0	2.6	3.0	Satisfactory

Table 3.26 Vibration values measured at the measuring point 2H, VS (2nd stage)

Output speed [rpm]	7.23	7.59	7.63	Valuation
HFD [g]	2.4	1.8	2.0	Good
ACC [g]	2.3	1.9	2.0	Good
ENV up to 10 kHz [gE]	2.2	1.9	1.9	Good

Table 3.27 Vibration values measured at the measuring point 2H, V (2nd stage)

Output speed [rpm]	8.96	8.82	8.62	Valuation
HFD [g]	2.8	2.1	2.2	Good
ACC [g]	2.8	2.2	2.3	Good
ENV up to 10 kHz [gE]	2.7	2.2	2.9	Good

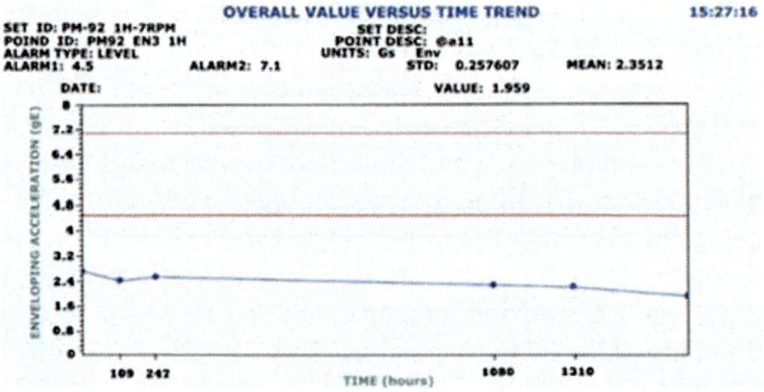

Fig. 3.67 Trend graph—summing values of high-frequency vibrations in the first measuring point

The measured values of the first measuring point show acceptable and stable condition, without signs of damage of the contact surfaces (Fig. 3.67).

In the second measuring point can be stated good condition, although the measured values show some changes—however, they are in standard of set limits (Fig. 3.68).

3.4.2.2 Analysis of the Lubricant Condition

When analyzing the condition of the gear, there was made a diagnosis of lubricant samples at the same time in order to determine:

- thermo-oxidative degradation of the oil and the contact surfaces, in particular in view of the formation of carbon, carbon black, and iron oxide of type of FeO, Fe_2O_3, and Fe_3O_4
- Fe wear particles (quantity—number, size, and type)

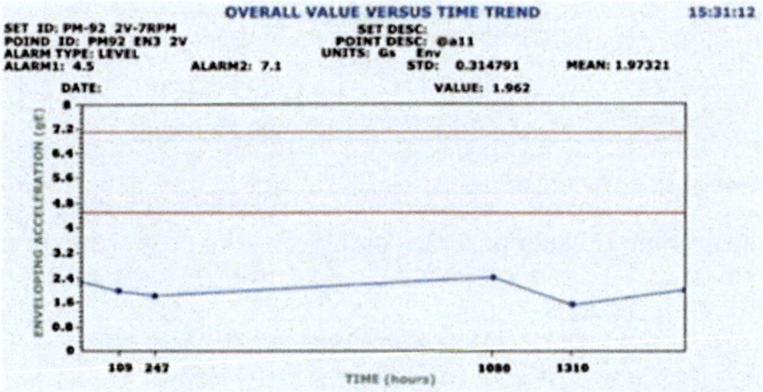

Fig. 3.68 Trend graph—summing values of high-frequency vibrations in the second measuring point

A summary of the measured values for individual samples of lubricant for monitored parameters is given in Table 3.28.

Measurement of total impurities is performed by usage of a special device made just for this purpose. A measuring instrument was used as CCT infra device, which works on photometric principle. On the device, there is a scale which directly shows the (%) of total impurities. By measuring the wear particles of WPC, it is determined the amount of large particles Dl and multitude of small particles Ds. The ideal condition occurs when Dl < Ds. For measuring WPC, a device—(direct-reading) capillary magnetic analyzer PMA 87-2 by which the sample passes through the micro tube which is stored in the magnetic field—is used. Due to the impact of this field, the wear particles are deposited in the capillaries. From the digital display of this device, it can be directly read the number of Dl and Ds. From these values, we can calculate

Table 3.28 The values of monitored parameters of chosen samples of lubricant

Sample number	Number of hours in operation [h]		Output speed [rpm]	Parameters monitored			Valuation of the oil sample condition
	Oil fill	Gearbox		**CCT IR** [%]	**WPC**[%]	**PLP** [%]	
1	11 1. PP-90	11	7.23 8.96	–	–	–	
2	109 1. PP-90	109	7.23 8.96	**Over 2.0**	**96.6**	**39.5**	Nonconforming **A2—DANGER**
3	138 2. X-220	247	7.59 8.82	1.1	15.8	18	Satisfactory
4	399 2. X-220	508	7.63 8.62	1.5	17.7	18.5	Satisfactory
5	638 2. X-220	747	7.63 8.62	1.6	18.3	19.6	Satisfactory
6	841 2. X-220	950	7.63 8.62	1.6	19.8	19.3	Satisfactory

Table 3.29 The recommended limits for the content of wear particles

Limits for the content of wear particles in gear oil		
ALARM 1	300 Fe (ppm)	Alert
ALARM 2	**400 Fe (ppm)**	**Danger**

the WPC—the amount of wear particles and PLP—the proportion of macro-particles in the WPC (%).

$$WPC = \frac{\mathrm{Dl} \pm \mathrm{Ds}}{1800} \times 100(\%) \tag{3.16}$$

$$PLP = \frac{\mathrm{Dl} \times \mathrm{k}}{\mathrm{Dl} + \mathrm{Ds}} \times 100(\%) \tag{3.17}$$

where k is the coefficient of difficulty.

The alleged state of exceeding the recommended limits (Alarm 2—Table 3.29) was confirmed even after completing the necessary analyzes of ferrograph photos—wear particles (Fig. 3.69) and for all basal monitored parameters (thermal, oxidative degradation of oil, the amount of solid wear particles) in the first sample of lubricant (Madit PP-90)—Table 3.30, after running-in, i.e., 109 h of operation.

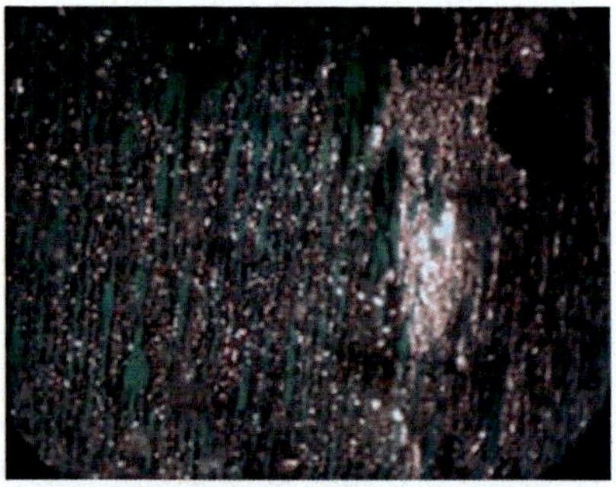

Fig. 3.69 Photo of ferographic traces of oil sample Madit PP-90 after running-in

Table 3.30 Content value of the solid wear particles for selected samples of lubricant

Sample number	Oil filling number	Lubricant type	Number of hours in operation [h]		Fe content (ppm)
			Gearbox	Oil	
1	1	PP-90	11	11	1800
2	1	PP-90	109	109	2710
3	2	X-220	247	138	140
4	2	X-220	508	399	150
5	2	X-220	747	638	160
6	2	X-220	950	841	170

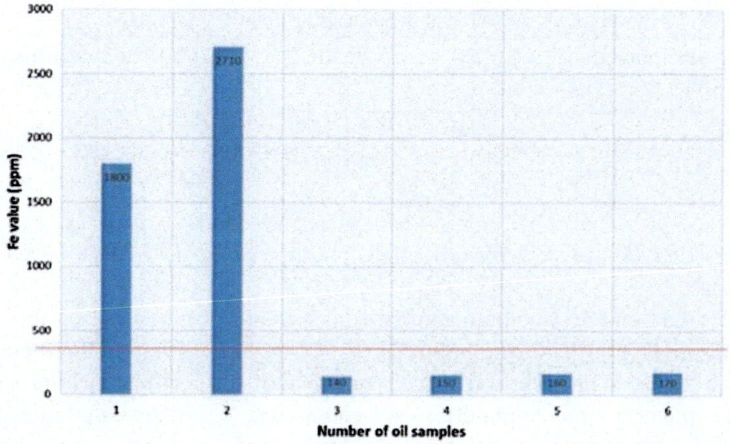

Fig. 3.70 Trend analysis of monitoring the content of Fe particles in a gear oil during the course of the test

After changing the oil, the condition of lubricant and friction mode was good (see Table 3.30). All monitored parameters, especially oil degradation and tested contact surfaces, were well below the recommended limit.

Graph (Fig. 3.70) shows the trend analysis, monitoring of Fe particles in a gear oil during the course of the test. The results confirm previous analyzes, in which after oil change there was a significant improvement in the operational status of the system under review.

3.4.2.3 The Course of Monitored Parameters—Summing Values—Part B

Terms of measurements taken in the following analysis are summarized in Table 3.23. In Tables 3.31 and 3.32, there are comparisons defined—trends of measured summing values of vibration.

The measurement was carried out at the place of 1 H, V—tooth meshing on the first stage. After the initial running-in of surfaces, the improvement occurred—a decrease in monitored parameters. All monitored parameters of high-frequency vibration show

Table 3.31 Vibration values measured at the measuring point 1H (1st stage)

Output speed [rpm]	5.0	5.0	5.36	Valuation
HFD [g]	2.1	1.5	1.9	Good
ACC [g]	1.8	1.4	1.9	Good
ENV up to 10 kHz [gE]	2.3	2.2	2.0	Good

Table 3.32 Vibration values measured at the measuring point 1 V (1st stage)

Output speed [rpm]	5.0	5.0	5.36	Valuation
HFD [g]	2.1	1.4	1.8	Good
ACC [g]	2.0	1.3	1.9	Good
ENV up to 10 kHz [gE]	2.3	1.3	2.0	Good

good, respectively satisfactory, operating condition of the teeth meshing and bearing conditions.

Tables 3.33 and 3.34 contain comparisons—trends of measured summing values of vibration. It is a measuring point 2H, V—of tooth meshing on the second stage. Even the second stage after the initial running-in of surfaces shows improvement—decrease in monitored parameters. All monitored parameters of high-frequency vibration demonstrate good operating condition, tooth meshing, and good condition of bearings.

In the first measuring point, the measured values show a running-in condition over the limit—danger, but after some time the condition stabilizes in good operating mode (Fig. 3.71).

The measured values of the second measuring point are in the running-in mode stable, later in poor condition (Fig. 3.72).

Acceleration enveloping measuring (gE, PtP, up to 10 kHz), Fig. 3.73—FFT spectrum presents the values in the required limit, or with a slight increase, which could be caused by running-in process. Time record shows an increase in the values compared with the required status, which highlights the deterioration, symptom of metal contact, and incomplete oil film—lubricant (Fig. 3.74).

Table 3.33 Vibration values measured at the measuring point 2H (2nd stage)

Output speed [rpm]	5.0	5.0	5.36	Valuation
HFD [g]	2.6	2.4	3.6	Satisfactory
ACC [g]	1.8	1.4	1.8	Good
ENV up to 10 kHz [gE]	1.8	1.1	1.8	Good

Table 3.34 Vibration values measured at the measuring point 2 V (2nd stage)

Output speed [rpm]	5.0	5.0	5.36	Valuation
HFD [g]	1.9	1.3	2.0	Good
ACC [g]	2.3	1.4	1.8	Good
ENV up to 10 kHz [gE]	2.4	1.5	2.0	Good

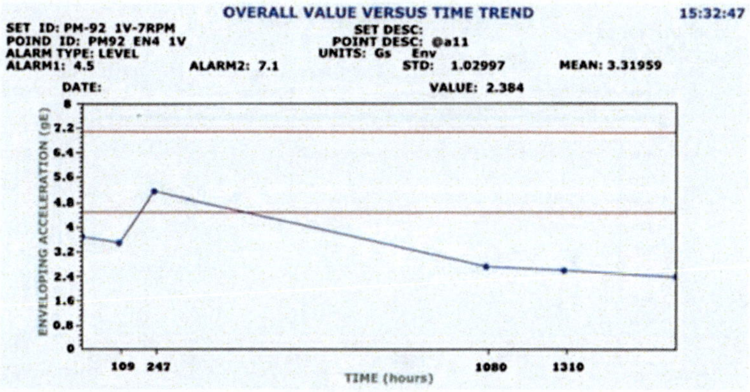

Fig. 3.71 Trend graph—summing values of high-frequency vibrations in the first measuring point

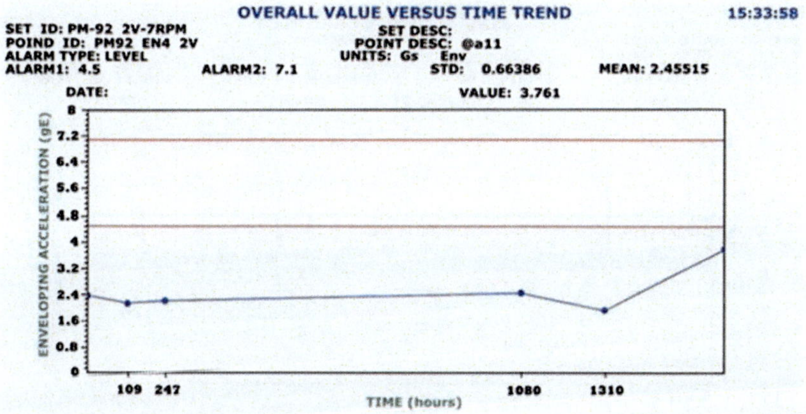

Fig. 3.72 Trend graph—summing values of high-frequency vibrations in the second measuring point

3.4.2.4 Analysis of the Lubricant Condition

During the tests of the gearbox carried out, lubricant has been analyzed in order to determine:

- thermo-oxidative degradation of the oil and the contact surfaces, in particular in view of the formation of carbon, carbon black and iron oxide of type of FeO, Fe_2O_3, and Fe_3O_4,
- Fe wear particles (quantity—number, size, and type).

Table 3.35 provides an overview of the conditions by collection of individual samples of lubricant and values of monitored parameters.

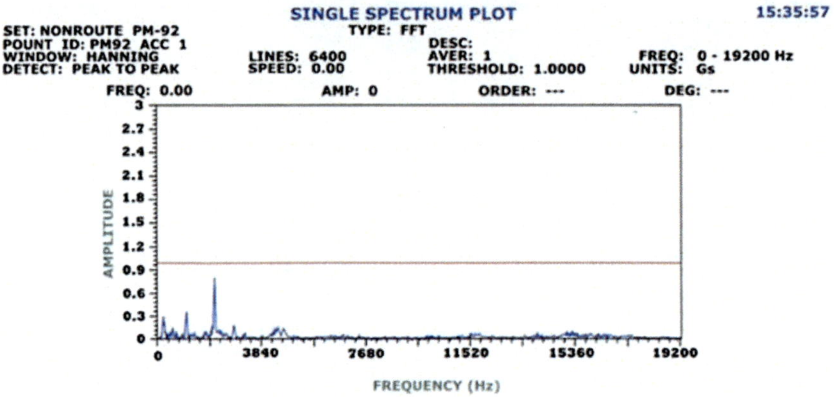

Fig. 3.73 FFT spectrum

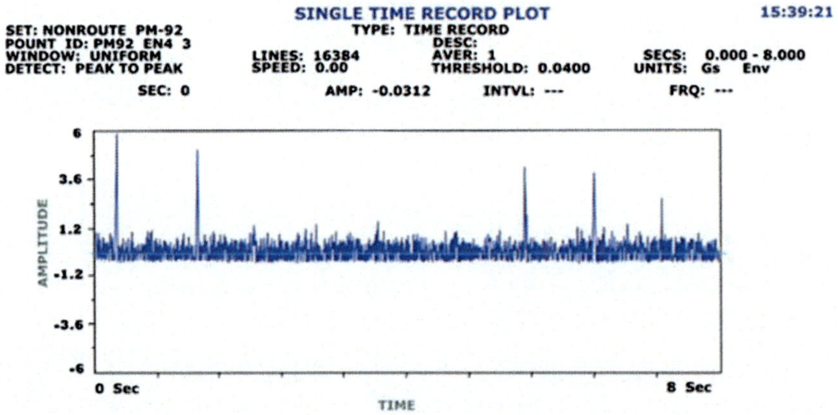

Fig. 3.74 TIME record

Table 3.36 contain the values of solid wear particles for selected samples of lubricant 1–5. Figure 3.75 shows a graph of trend analysis showing the Fe particles content in the gear oil during the course of the test for the second part of the analysis. The results confirm previous analyzes.

3.4.3 Results and Discussion

The first type of analysis—test, part A—applied a trend diagnostic measurement of the vibrations and acoustic emission, the assessment of the operating condition of the bearings and gearings, and assessment of the mode of friction with analysis of Fe

Table 3.35 Values of monitored parameters in selected samples of lubricant

Sample number	Number of hours in operation [h]		Output speed [rpm]	Monitored parameters			Valuation of oil sample condition
	Oil filling	Gearbox		**CCT IR** [%]	**WPC** [%]	**PLP** [%]	
1	30 1. X-220	30	5.0	0.6	15.8	18.0	Satisfactory
2	300 1. X-220	300	5.0	0.65	15.5	17.5	Satisfactory
3	500 1. X-220	500	5.0	0.74	15.0	17.5	Satisfactory
4	755 1. X-220	755	5.0	1.25	16.3	18.4	Satisfactory
5	1005 1. X-220	1005	5.36	1.30	17.0	19.0	Satisfactory

Table 3.36 The values of solid wear particles for selected samples of lubricant 1–5

Sample number	Oil filling number	Lubricant type	Number of hours in operation [h]		Fe content (ppm)
			Gearbox	Oil	
1	1	X-220	30	30	410
2	1	X-220	300	300	170
3	1	X-220	500	500	290
4	1	X-220	755	755	290
5	1	X-220	1005	1005	280

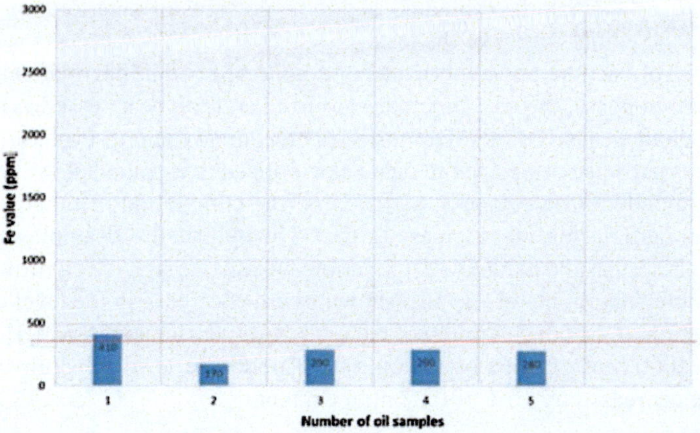

Fig. 3.75 Trend analysis of monitoring the content of Fe particles in a gear oil during the course of the test for the second part of the test

wear particles for two types of gear oil: Madit PP-90 and X-220, which are suitable for the lubrication of industrial gears, synchronized manual switches, clutches of differentials, axles, bearings, etc. On the basis of the measurements and analyses taken, it can be stated:

- In terms of measuring of high-frequency vibrations: Condition of contact surfaces showed moderate running-in processes moderate in terms of intensity in the early hours. After completion of running-in-smoothing of surfaces, all dynamic parameters monitored have been low; the measured signal was far below the alarms.
- In terms of friction mode, lubricant condition: The first oil filling showed unsatisfactory friction mode; all parameters monitored were high above the recommended limit. Particularly high was the content of Fe wear particles, approximately 9× more than the recommended guide value for A1-alert. The second oil filling showed significantly better condition in terms of thermal—oxidative degradation (of oil and surfaces) and also in terms of the amount of Fe wear particles. Even after 840 h of operation, all monitored parameters were in values of normative limits.

In the second part of analysis—test, part B—the trend diagnostic measurements of vibrations, assessment of the operational condition of bearings and gearings and assessment of the friction mode with the analysis of wear particles of type Fe for one type of lubricant were carried out. Summary of the results and analysis:

- In terms of measuring of high-frequency vibrations, it was confirmed that condition of contact surfaces showed running-in processes during the first hours, particularly at the input of central gearing (1st stage). After completion of running-in-smoothing of surfaces, all dynamic parameters monitored have been low; the measured signal was far below the alarms. The operating state of monitored contact surfaces can be evaluated as good—no signs of deterioration or damage during the whole course of the test.
- In terms of friction mode, lubricant condition: After an initial relatively moderate running-in phase, the friction mode stabilized and can be assessed as satisfactory throughout the next course of the test. After the initial running-in phase, the content of wear particles of Fe-type in gear oil stabilized at a value below 300 ppm (the recommended limit Alarm 1-Alert). Until the completion of the test, the levels of iron particles did not increase further. The collected oil samples showed low level of thermal and oxidative degradation. Distinctive sign—a confirmation of the favorable friction mode—is also shown by low decrease in the level of additives (depletion of the concentration of the main additive elements—Mo, P, S, and Ca). After 1000 h of gearbox operation in test mode, the level of additives in the oil filling decreased by 10% of the initial content.

In view of the high-frequency vibration measurements in both cases (analysis A and analysis B), at the end of a run-smooth surface it confirmed that all monitored dynamic parameters were below the level of set alarms. In view of the state of lubrication and friction in both cases, it confirms that the second load of transmission oil exhibited a much better shape both in terms of thermal–oxidative degradation (oil and surface) and of the quantity of wear particles of Fe.

The mechanisms for formation of the particles are relatively complex tribological processes; the particles are mostly products of adhesive and abrasive wear. Adhesive wear occurs when between the mutually moving surfaces there is not sufficient oil film, which is due to either a low viscosity oil or inadequate lubrication. Abrasive wear occurs most often when the abrasive particle gets between two mutually moving surfaces. The result of abrasive wear is a creation other particles that contaminate the oil [19].

Realized experiments and measurements have been shown to influence the type of transmission oil to the operating characteristics of planetary gear machining center. For example, the oil with lower viscosity may have a problem with maintaining the lubricant film, which clearly leads to greater wear. A suitable viscosity is always ultimately manifested in increased heating of transmissions, caused by either excessive friction between components or by difficult running of gear shifting mechanism. Of course, a huge impact on the properties of lubricants and oils is their type and composition of individual additives.

3.5 The Parameter Correlation of Acoustic Emission and High-Frequency Vibrations in the Assessment Process of the Operating State of the Technical System

In the process of technical diagnosis, a variety of techniques are applied to diagnose machines but it often happens that the measurements are not interpreted or interpretation is not correct. Within the frame of vibration diagnostics, there are occurrences of realizing the very collecting and recording of the vibrations signals while ignoring the force and energy that causes fatigue damage to machine parts and thusly shortens their lifespan. It means that twice the force shortens durability of the bearings to 1/8 of technical serviceability and device's lifespan is reduced from 5 to 10 years to only several months. Bearing diagnostics is a very demanding field and if technical diagnostician relies only on "their" method, the results of their work can even be fatal. Bearing damage can be caused by unsuiTable. lubricant, contaminated lubricant, water, excessive force, oxidation and degradation of the lubricant additives. The human factor can cause that bearing is installed with a larger overlap, overheated during installation or it is over-lubricated, insufficiently lubricated or without oil. These difficulties can shorten the bearing lifespan by a few hours up to several months [2, 20].

Based on cooperation of our department with practice the demand arose for accurate assessment of the operating condition of gearboxes. Four selected gearboxes were analyzed. In the context of diagnostic, it was applied measurement of high-frequency vibration and acoustic emission. The measured values were used for the cross-correlation and verification of selected methods.

3.5.1 Description of the Measuring—Characteristics of the Machines and Measuring Methods

In the process of diagnostic activities for the purpose of vibration measurement, process data collection, and subsequent analysis, the following diagnostics systems were used:

- the frequency analyzer and Data Collector Microlog GX,
- the software interface of Aptitude Analyst,
- vibration sensor, accelerometer Wilcoxon Research Model SKF786M, sensitivity 100 mV/g, frequency range from 1 to 20 000 Hz.

Measuring method based on the recommendations of the technical standard STN ISO 10816-3—Mechanical oscillation—was used. Evaluation of machine oscillation by measuring the non-rotating parts of machine is done.

For the purpose of vibration assessment in the low-frequency area, the following were applied:

- The measurement method, velocity (vibration velocity); unit, mm/s; and RMS detection.
- According to the recommendations of ISO 10816–3, FFT (fast Fourier transform) frequency spectrum for the area from 10 to 800 Hz.
- The measurement method, enveloping acceleration (method of measuring the vibration acceleration in the high-frequency range to 10–20 kHz through which assessment is carried out in particular for the state of the rolling bearings and gear ratio), unit Eg, detection of PtP, the frequency area from 50 to 1000 Hz, FFT spectrum, and time record.

For the assessment of vibration in the high-frequency area, the following methods were used:

- The measurement method, acceleration (vibration acceleration); unit, g; detection of PtP, frequency range up to 16 kHz, the FFT spectrum, and time record.
- The measurement method, acceleration enveloping; unit, Eg; detection of PTP; frequency range up to 10–20 kHz; and the FFT spectrum, and time.

Diagnostic methods for the assessment of acoustic emission and ultrasound:

- The measurement method, See (Spectral Emitted Energy)—acoustic emissions (SKF), unit See, detection of PtP, frequency range up to 600 kHz, the FFT spectrum, and time record.
- The measurement method HFD—high-frequency detection method (SKF), analysis of vibration (acceleration) in frequencies 40 kHz.
- Method for assessment of the state of lubrication, resistance of the oil film, metallic contact, sticking, and abrasion of contact surfaces (bearings and gear).

3.5.1.1 Recording and Running of Operating Parameters—Realization of the Experimental Measurements

Table 3.37 contains a summary of the measured values of the exposure of noise and vibrations of the individual gearboxes marked A, B, C, and D (Figs. 3.76 and 3.77).

The measured acceleration values were obtained by diagnostic system with the application of accelerometer Wilcoxon Re-search, SKF786M in cooperation with the PorTable. Data collector/FFT Analyzer SKF Microlog Analyzer (Fig. 3.78). The gearboxes were measured at one measuring point (acceleration).

Table 3.37 Measured exposure values for noise and vibrations of gearboxes A, B, C, and D

Gearbox	HF vibrations—oscillations			Sound—noise			Evaluation
	Acc	Acc	EnvAcc	LAeq	LASmax	LAImax	
	[g]	[g]	[gE]	[dB]	[dB]	[dB]	
	To 20 kHz	Zone 1–6 kHz	To 10 kHz				
A	8.09	6.87	3.67	79.5	81.0	84.6	A2
B	5.30	4.49	2.45	73.6	74.7	78.0	A2
C	4.43	3.07	1.58	72.5	73.8	75.5	A2
D	*3.20*	*1.37*	*0.92*	*72.4*	*73.5*	*74.6*	*SUITABLE*
ALARM 1 (A1)	3.30	2.20	1.00	–	–	–	
ALARM 2 (A2)	4.30	3.00	1.50	75.0	75.0	75.0	

Explanatory notes: Acc—acceleration of vibration displacement, EnvAcc—envelope of vibration acceleration, Laeq—value of the equivalent level of acoustic pressure A, LASmax—the maximum level of acoustic pressure using the weighting filter type A and the time constant "Slow," LAImax—the maximum level of acoustic pressure using the weighting filter type A and the time constant "Impulse."

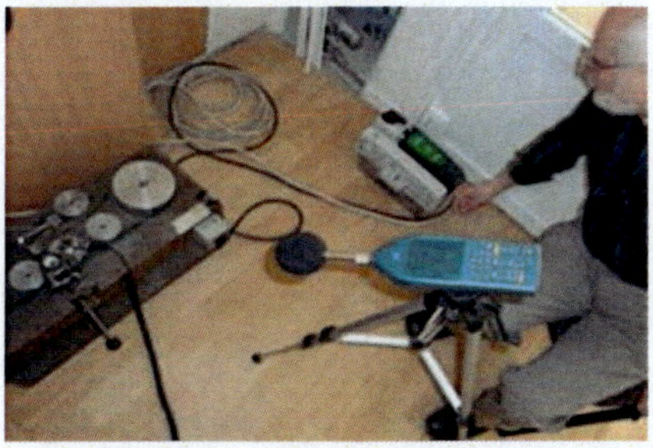

Fig. 3.76 The sound pressure measurement

Fig. 3.77 The contactless measuring of vibration by the laser vibrometer

Fig. 3.78 Accelerometer model Wilcoxon Research, SKF786M and PorTable. Data Collector/FFT Analyzer SKF Microlog Analyzer

The values of alarm levels (primarily for noise limits, sound pressure, acoustic emission, and high-frequency acceleration) are determined by the following ways:

- by agreement between a supplier (manufacturer) and customers,
- when the signal increase over 200% compared to the reference (agreed) level measured on the normal product quality and design,
- by comparing the dynamic characteristics of competing manufacturers–suppliers), respectively, by comparing the measured values with the normal level of the same products in the local market.

In the specific case of defined alarm, set points were based on an analysis of signal levels of 95% of the measured values of controlled amount of the same types of mechanisms (see Table 3.37).

From the diagram of the measured values of the gearbox "A" (Fig. 3.79), it is seen that the value of the frequency about 2700 Hz achieved the level of noise 81 dB, respectively, 84.6 dB.

FFT spectrum of a vibration signal confirmed. the statement as with value about 2700 Hz measured signal value was: EnvAcc—3.67 (gE), Acc—6.87 (g), Acc—8.09 (g). Time record of signal points to the unsatisfactory operating condition of the gearbox (Fig. 3.80). In both cases, the measured values are set above the limit of alarm A2 (see Table 3.37).

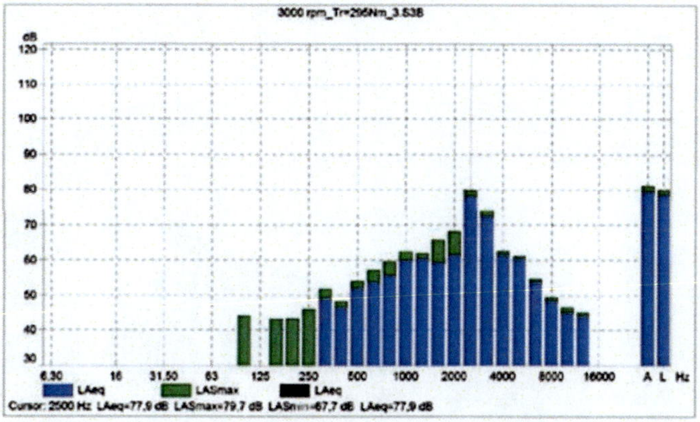

Fig. 3.79 Graphic record of acoustic emission for gearbox "A"

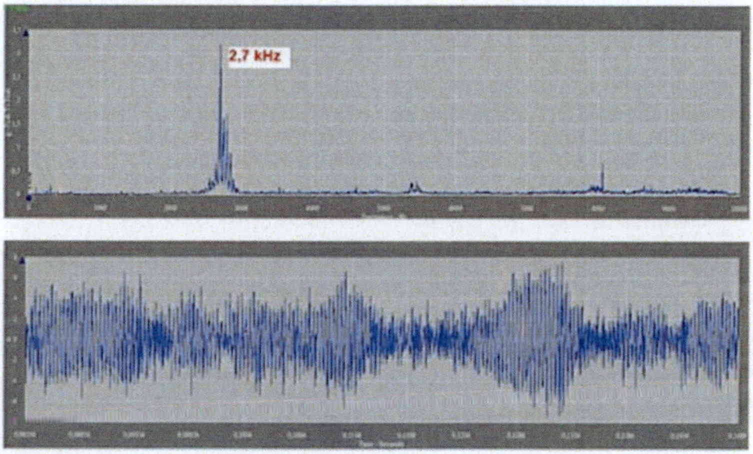

Fig. 3.80 FFT spectrum and recording TIME of gearbox "A"

In the case of the gearbox marked by the symbol "B," it is possible to record the state from the diagram when the vibration value about 4000 Hz is equal to the noise level of 74.7 and 78 dB (Fig. 3.81).

FFT spectrum of a vibration signal confirms the state where with value about 4200 Hz the measured signal value was: EnvAcc—2.45 (gE), Acc—4.49 (g), Acc—5.30 (g). Time record of signal points to the unsatisfactory operating condition of the gearbox "B" (Fig. 3.82). In both cases, the measured values are set above the limit of alarm A2.

For the third case—gearbox "C" in the vibration value about 4000 Hz had a vibration sound emission moving at the level of 73.8 and 75.5 dB (Fig. 3.83). FFT spectrum of a vibration signal confirmed that with value around 4150 Hz the measured

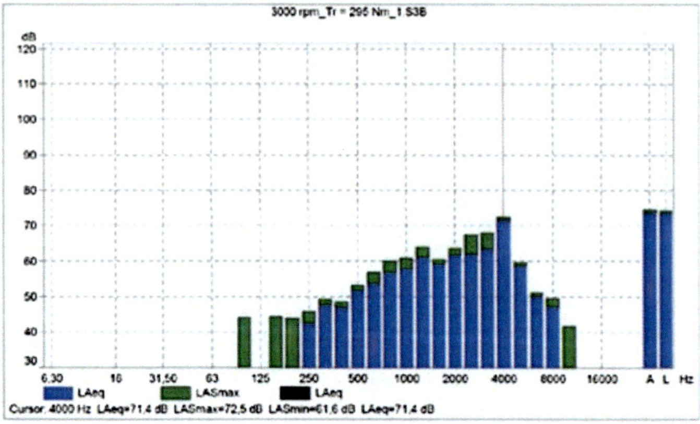

Fig. 3.81 The graphic record of acoustic emission for gearbox "B"

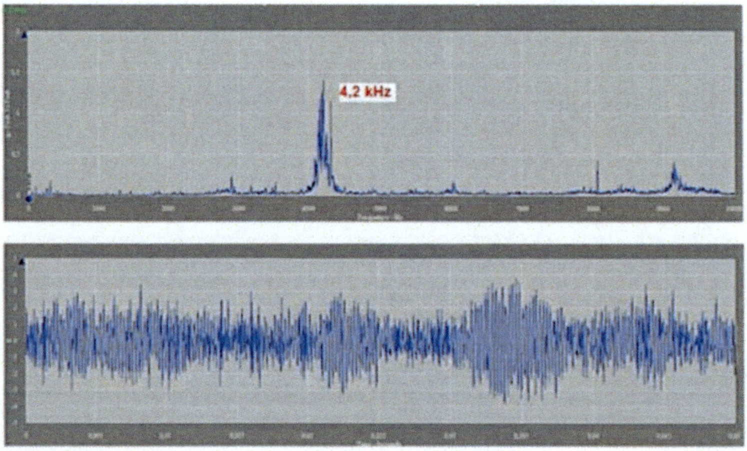

Fig. 3.82 FFT spectrum and recording TIME of gearbox "B"

signals value was: EnvAcc—1.58 (gE), Acc—3.07 (g), Acc—4.43 (g). Time record of signal points to the unsatisfactory operating condition of the gearbox "C" (Fig. 3.84). In both cases, the measured values are set above the limit of alarm A2.

In the case of the last gearbox marked by the symbol "D" was recorded a state where the all zones were located below level of 70 dB without dominant frequencies across the whole monitored zone (Fig. 3.85).

FFT spectrum of the vibration signal also confirms that within measurement area, there is not a state of increased signal. The measured signal value was: EnvAcc—0.92 (gE), Acc—1.37 (g), Acc—3.20 (g). Time record points to the satisfactory operating condition of the gearbox (Fig. 3.86). The measured values of both cases are within the defined limits, i.e., condition is satisfactory.

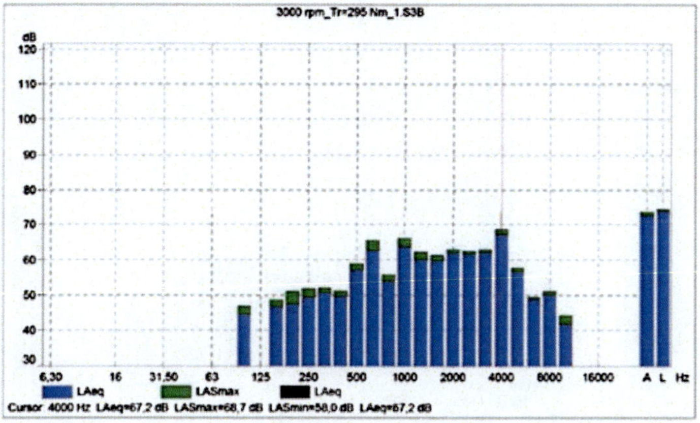

Fig. 3.83 The graphic record of acoustic emission for gearbox "C"

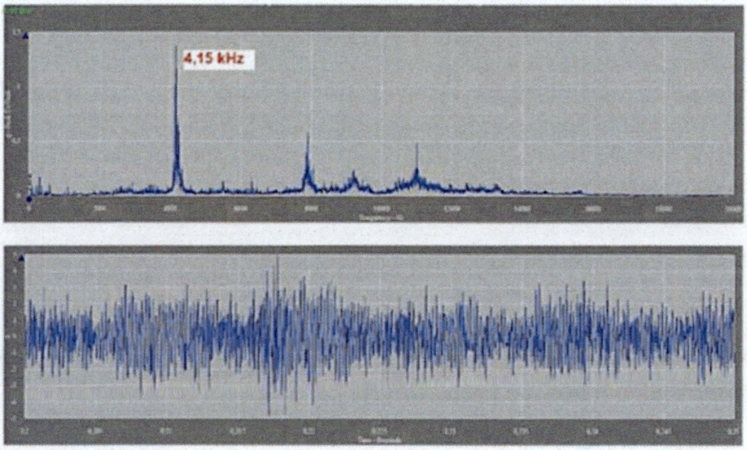

Fig. 3.84 FFT spectrum and recording TIME of gearbox "C"

3.5.1.2 Discussion of the Measurements Implemented

Noise and vibrations are closely related since in many cases the oscillating bodies are the cause of secondarily created noise and vice versa—air-transmitted acoustic energy can create interfering oscillation of constructions, worsening of workplace conditions, etc.

The causes of vibration in technical practice are alternate forces that arise from the clearance and inaccuracies in rotating movements of machines, in vehicles movement and their parts, at work with the various machine tools, etc. Primary task of vibration diagnostics is early identification of moment of required maintenance or possibly changing the segment in assessed system before end of its working life. This type

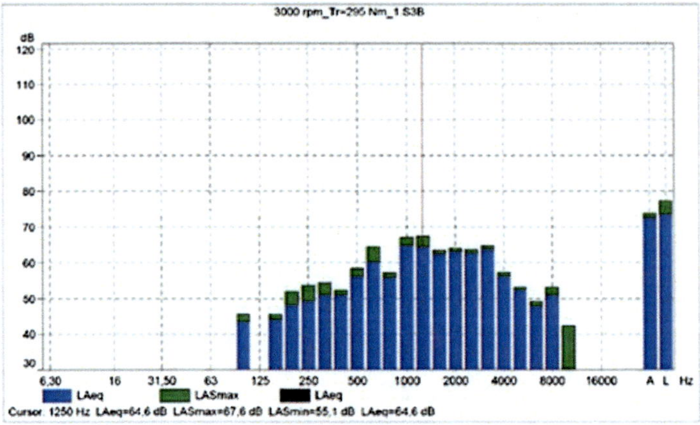

Fig. 3.85 Graphic record of acoustic emission for gearbox "D"

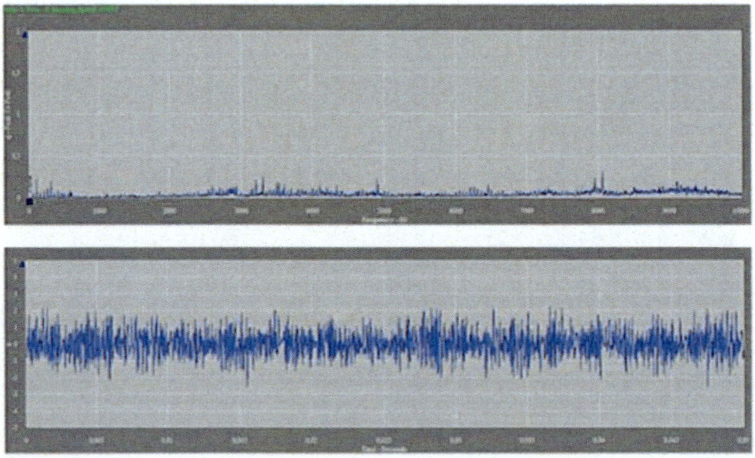

Fig. 3.86 FFT spectrum and a time record of gearbox labeled "D"

of diagnostics increases security, reliability, and durability, the period between two repairs, respectively, finally thereby reducing maintenance costs.

In respect to case of diagnostics of assessed precision gear-boxes described in the study, it is possible based on realized measurements and analysis of dynamic signal by means of correlation of two methods, to state the following conclusion:

- operating state of gearboxes "A," "B," and "C" is unsatisfactory. The measured values are identified above the recommended limit of warning Alarm 2 (A2—danger),
- gearbox "D" is in satisfactory working condition for all measured parameters and measurement methods.

After the realized repair of gearboxes "A" to "C," it is recommended to rediagnose their operating status by measuring high-frequency vibrations. Consequently, it is desirable, by measuring the microgeometry of contact surfaces, to assign the measured value of high dynamic signal for specific gearbox component and define its eventual production inaccuracy.

References

1. Baron, P. - Kočiško, M. - Hlavatá, S. - Franas, E.: Vibrodiagnostics as a predictive maintenance tool in the operation of turbo generators of a small hydropower - 2022.In: Advances in Mechanical Engineering. - Thousand Oaks (USA) : SAGE Publications vol. 14, no. 5 (2022), pp. 1–15 [online, print]. - https://doi.org/10.1177/16878132221101023.
2. Valent, O. - Galád, M. and Kačmár, Ľ.: Technická diagnostika 1, CMMS, Praha, Czech Republic, 2010.
3. Vojtko, I. - Baron, P. - Pollák, M. - Kaščak, J.: Examining the effect of alignment of the rotor of the emissions exhaust fan on Its operating parameters - 2019.In: Advances in Materials Science and Engineering. (2019), pp. 1–13 [print]. - ISSN 1687–8434
4. Ballo, I., Skúšobné otázky pre vibrodiagnostiku, Nitra, COPTD, 2001.
5. Juliš, K. - Borůvka, V. - Fryml, B., Základy dynamického vyvažování. SNTL 1979.
6. SKF Ložiska, a.s., Praha. Ustavování hřídelů, *Aplikační materiál FIXTUR LASER*
7. Piotrowski, J.: Shaft Alignment Handbook, CRC Press Taylor and Francis Group, 2007, ISBN-13: 978-1-57444-721-7, 865 p.
8. Šeremeta L., Podkladové materiály k ustavování, LAMI KAPPA s.r.o., Teplice, 2004
9. Kuric, I. - Císar, M. - Tlach, V. - Zajačko, I. - Gál, T. - Więcek, D.: Technical diagnostics at the department of automation and production systems, 2nd International Conference on Intelligent Systems in Production Engineering and Maintenance, ISPEM 2018; Wroclaw; Poland, Volume 835, 2019, Pages 474–484
10. Murčinková, Z. et al., Vibration and Temperature Decreasing Through the Material Damping and Tool Path Strategies Applied for Milling the Difficult-to-machine Materials, Advances in Research, 2015, 4: 269–278. ISSN 2348-0394
11. Asiri, S.: Tunable Mechanical Filter for Longitudinal Vibrations, Shock and Vibration, 2007, 14: 377–391. ISSN 1070-9622
12. Panda, A. - Prislupčák, M.: Valuation of vibrations during machining, Trends and Innovative Approaches in Business Processes, 17. International Scientific Conference, Almanac, Košice, 2014, pp. 1–5.
13. Baron, P. - Kočiško, M. - Blaško, L. - Szentivanyi, P.: Verification of the operating condition of stationary industrial gearbox through analysis of dynamic signal, measured on the pinion bearing housing - 2017.In: Measurement. Vol. 96 (2017), p. 24–33. - ISSN 0263-
14. Salokyová, Š. - Ružbarský, J. - Krenický, T.: Vibration diagnostics in manufacturing system with water jet technology, RAM-Verlag Lüdenscheid, Germany, 2015, 118 p.
15. Palaščáková, D. - Svetlík, J.: A new approach to manufacturing strategy, Global management and economics. Vol. 2015, no. 1, p. 91–95, 2015
16. Sinay, J. - Pačaiová, H. - Tomašková, M. -Kopas, M.: Application of technical risk theory for evaluation of gearboxes damaging processes, IEA 097. - Helsinki : Finnish Institute of Occupational Health, 1997, p. 560–565, 1997
17. Baron, P. - Kočiško, M. - Dobránsky, J.: The dynamic parameters correlation assessment of the textile machine high-speed bearings in changed technological conditions - 2018.In: Measurement. Vol. 116 (2018), p. 575–585. - ISSN 0263-2241
18. Adamčík, P. : Systems for diagnostics of machinery, AT&P Journal 5 (2005), p. 20–22

19. Baron, P. - Kočiško, M. - Dobránsky, J. - Pollák, M. - Töröková, M.: Research and application of methods of technical diagnostics for the verification of the design node - 2016.In: Measurement. Vol. 94 (2016), p. 245–253. - ISSN 0263-2241
20. Baron, P. - Dobránsky, J. - Kočiško, M. - Pollák, M. - Cmorej, T.: The parameter correlation of acoustic emission and high-frequency vibrations in the assessment process of the operating state of the technical system - 2016. In: Acta Mechanica et Automatica. Vol. 10, no. 2 (2016), p. 112–116. - ISSN 1898-4088

Chapter 4
Tribotechnical Diagnostics

Tribotechnical diagnostics is one of the methods of technical diagnostics without disassembly, which utilizes lubricants as media for obtaining information. Its task is to identify, evaluate, and report the occurrence of foreign matter in the lubricant. Appropriate interpretation of the results from the performed experiments allows not only for a timely notification on emerging failures, but also for the localization of the place of formation of mechanical error [1–4].

The decisive factor for tribotechnical diagnostics is the speed and accuracy of performed analysis. A variety of methods with different degrees of usability has been developed for the assessment. Individual types can be divided into three groups [1, 5]:

- Simple tests for operational inspection of lubricants.
- Standard tests for accurate determination of the lubricants' quality.
- Specific methods for the overall diagnosis of lubricants and machinery equipment.

Consideration of the operational reliability of the machinery or its drive unit with use of the tribotechnical methods brings a considerable economical savings in the operation. Tribotechnical elements are components of larger mechanical systems, and therefore, the interaction between elements of tribological system and other components is decisive for its behavior [1, 6].

The method of detection and isolation of particles using the magnetic analyzer allows detecting the intensity of wear and clarifies the process of friction and wear. It is the diagnostic method using the particles emerging from fraction for the identification of the character of the tribological processes at friction surfaces. It allows the differentiation of the particles into size groups. The principle of the ferrography lies in the separation of the particles according to their size in strong magnetic field. It is used for the assessment of wear intensity on friction points of the aggregate. The sample of considered oil flows down the slant pad situated in magnetic field. The biggest particles are sedimented at the beginning of the transparent pad while another particles progressively according to their size and shape. Important is also

P. Baron et al., *Application of Troubleshooting Tools in the Monitored Production Processes*, Management and Industrial Engineering,
https://doi.org/10.1007/978-3-031-41428-2_4

the orientation on the ferrogram, the surface and the shape of the particles. Obtained ferrogram is then evaluated using the microscope. It can be evaluated in two ways—ferroscopic one (determining the morphology and the chemical constitution of the particles)—and ferrodenzimetric one (evaluating the distribution of the particles in consideration with their size).

The principle of the ferrography with direct reading remains the same as in the above mentioned method, though the glass slide is replaced here with thin walled glass capillary fixed in divergent magnetic field of a special magnet. The tested oil flows through the capillary. In two places, where the large and small particles settle, there are fixed radiation sources and sensors, which sense the decrease in radiation intensity. It is directly proportional to the degree of coverage of the capillary wall with particles.

4.1 Classification of Lubricants

Lubricants as intermediates deliberately used in tribological systems are essential for the operation of machines and equipment and, when properly used, guarantee the safety, reliability, economy and designed life of the friction nodes, and thus of the entire equipment during its operation in technical deployment. In order for lubricants to meet the requirements that are imposed on them and are expected from their function, they must have certain tribochemical and tribophysical properties.

The basic property of the lubricant is its lubricating ability, that is, the property of transferring forces in the friction node by internal friction while creating, if possible, continuous layers on the surface of the friction bodies.

The requirements placed on the lubricants are [7, 8]:

- Reduction of mechanical energy losses and improvement of the mechanical efficiency of the system,
- Reduction or suppression of wear and tear and its harmful effects on the tribological system,
- Protection against dirt that can enter the system from the outside and thus also the ability to get rid of friction surfaces created by dirt,
- Protection against corrosion and chemical neutrality,
- Sufficient cooling and better heat dissipation,
- Lubricant functioning as a sealant.

From the chemical point of view, they can be divided into [7, 8]:

- Organic,
- Inorganic.

According to their origin, the lubricants are:

- Natural,
- Synthetic.

The current assortment of lubricants is vast and they are divided according to their fitness for specific operating conditions and special purposes into [7]:

- Liquid,
- Plastic,
- Rigid,
- Gaseous.

Liquid lubricants

At present, these are the most common lubricant type. Their advantage is that they come in a wide range, allowing for the use in the entire area of liquid and mixed friction, i.e., in hydrodynamic, hydrostatic, elastohydrodynamic lubrication.

Liquid lubricants can be divided into:

- Chemically distinct substances (H2O, H2SO4, glycerin, certain synthetic oils),
- Homogeneous mixtures (consisting of molecules of the same chemical composition, but of different sizes. These include lubricating oils with a very wide range of viscosity),
- Liquid dispersions (particles of one substance dispersed in another or in several substances without blending together).

They can be further divided into [7, 8]:

I. Vegetable and animal oils

They belong to the oldest group of lubricants and they include rapeseed, castor, olive and bone oils. They share a common feature, which is adhesion.

A disadvantage of all these oils is that they age quickly. This is manifested by the sticky effects of resins and sediments.

II. Spirits

They are produced by the vacuum distillation of crude oil or brown coal tar. They are inexpensive oils, have a high sealing capacity and excellent adhesion. When free from asphalt substances during production, they are suitable for lubrication of bearings and sliding housings in agricultural machinery and rail transport.

III. Refined lubricating oils

They are obtained after processing of distillates mainly with acids, hydroxides or solvents. They have a high flash point, greater lubrication capacity and, above all, higher resistance to aging. These properties help to remove not only impurities, but also improve the hydrocarbon composition.

At present, refined lubricating oils are being outperformed by new production processes, especially high-pressure hydrogenation.

IV. Greasy oils

Their applicability is very small. They are interesting in that they exhibit lower surface tension compared to metals than refined lubricating oils. They are used to lubricate racing two-stroke engines.

V. Alloyed, additive oils

These are mineral oils refined with admixtures that we call additives. Additives are chemically very complicated substances and can be divided into:

- Organometallic, i.e., ash additives (currently little used),
- Organic, often polymeric.

VI. Synthetic oil

Synthetic oil is a lubricant consisting of chemical compounds that are artificially produced by synthesis from substances other than crude oil. These oils are often used instead of the lubricants obtained from crude oil refining because they offer better mechanical and chemical properties than traditional mineral oils.

Greases

They are colloidal systems of a subgroup of gels, less frequently sols, or jellied sols. Macroscopically, they are buttery, fibrous, spongy, or granular. They consist of a liquid phase formed by the lubricating oil and solid particles of the dispersed substance.

These lubricants are manufactured from petroleum oils which are fortified with metallic soaps and other substances to achieve appropriate texture. An important property is their drip point, which determines the temperature at which the grease transforms into a liquid state [7].

According to their stability, we can divide them into [7]:

- Thermally stable lubricants (the structure is not disrupted by the temperature or is regained after cooling),
- Mechanically stable lubricants (rheological properties differ little in disrupted state),
- Colloidal unstable lubricants (they break down into viscous phase and insoluble colloid—xerogel).

Depending on the type of hardener, which usually affects the properties of the lubricant more significantly than oils, we can divide them into [7]:

- Soaps,
- Non-saponified,
- Hydrocarbon,
- Halogenated.

In terms of their operational deployment, we classify greases as fit for [7]:

- Roller bearings
- Plain bearings,
- Gears,
- Sliding guides,
- Mobile equipment

Solid lubricants

They are used as a structural material in the manufacture of self-lubricating bearings, pin sockets, gears for self-lubricating wheels and material for the production of radial shaft seals. These lubricants can be used as solids with low hardness, low adhesion to the surfaces of solid friction bodies and a layered structure guaranteeing low shear strength. They can withstand high-pressure loads and do not change their physical properties at high temperatures [7].

Gaseous lubricants

They are a less common lubricant type. They are limited in use for aerostatic or aerodynamic lubrication of plain bearings and housings. Steel–steel and steel–bronze are used as common pairs of materials of sliding pairs of lubricated gases. It is convenient to use materials with good properties in emergency operation.

The above-mentioned housings use the properties of certain gases and their mixtures, such as [7]:

- Their ability to work in a radioactive environment,
- Small pressure losses in the housing,
- Chemical stability at high temperatures,
- Elimination of cavitation,
- etc.

A drawback of gaseous lubricants is mainly their instability and the formation of turbulent flow, high demands on the accuracy of their production and assembly of the sliding pair and the possibility of friction occurring in solid bodies when the conditions of aerostatic or aerodynamic lubrication are infringed [7, 8].

The oil must [7]:

- Resist shear force fields, e.g., even in the most difficult lubricated area—the upper dead center of combustion engines,
- Have good adhesion to the lubricating surface under all operating conditions,
- Have a good ability to dissipate friction and operating heat,
- Protect ferrous and non-ferrous metals from corrosion,
- Resist aging as much as possible, even under unfavorable working conditions—i.e., resist the oxidation of hydrocarbons or basic components of the oil,
- Ensure operation in severe frosts, as low as—40° C and less,
- Ensure that the impurities arising from abrasion and scorch are as fine as possible,
- Contribute to the sealing of the piston in the cylinder even at high temperatures and try to minimize the blow-through of gases.

The oil should be [7]:

- Resistant to evaporation, which is favorably reflected in small losses in operation at high and low temperatures,
- Of the shelf life spanning at least 2 years in the dark (cans, polystyrene),
- Economical in operation.

The oil must not [7]:

- Attack sealing materials,
- Foam in operation in the engine and other systems,
- Have a high carbonation number and must not be susceptible to the formation of cold sludge.

4.2 Research and Correlation of Diagnostic Methods for Assessment of the State of Oil Filling in Cycloid Gearbox

The objective of the realized analyses was to assess the condition of the oil filling in the cycloid gear of the manipulator (Fig. 4.1). The condition of the solid–abrasive particles in the oil was analyzed. The objective was to identify following relations:

- Run-in processes of contact surfaces, analysis of wear particles,
- Methods for the quantification of total content of Fe wear particles,
- Kinematic clearance (LM—lost motion), change after the run-in as a manifestation of the surface wear intensity,
- Correlation, quantification of measured parameters, LM and the Fe content of wear particles in the oil,

The tests were carried out with the focus on wear particles, which arise during the run-in or in the first hours of contact surfaces operation. At the same time, the correlation between total content of Fe particles in the oil and the increase in kinematic clearance in cycloidal gear was monitored.

Trial runs of the cycloid gearbox were in the frame of realized experiment observed from the viewpoint of the identification of abrasive particles during first 600 h of the

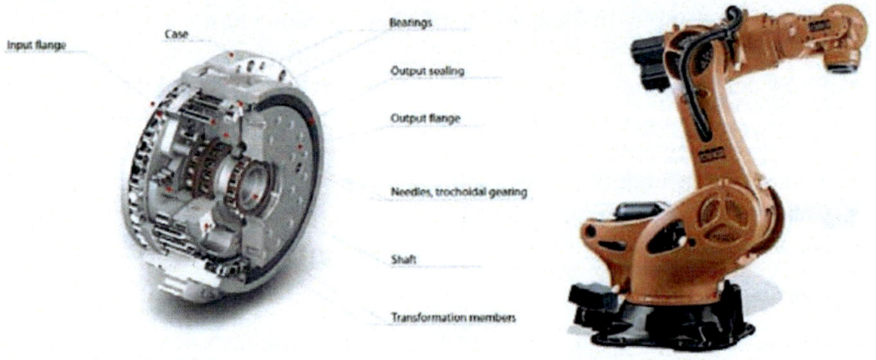

a.) b.)

Fig. 4.1 Diagnosed cycloid gearbox **a** applied in the joints of a robotic manipulator **b**

operation in five tested cycloid gear pairs (oil samples A, B, C, D, and E) of the gearboxes. We compared the application possibilities for two diagnostic methods—induction method (diagnostic device Cosmos—Steel Dust Checker) and x-ray fluorescent method, for identification of the friction particles on the Fe base. Method of the ferrography analysis was used for the verification and quality description of the friction particles (their type and size).

The principle of the x-ray fluorescent spectrometry is based on the emission of secondary (or fluorescent) x-ray of the material, which was irradiated by the high-energy x-ray or gamma radiation. Analyzed atoms of the sample are driven by generated x-rays. The electrons change their position in atomic shells and also their energetic level. These energetic differences are emitted in the form of special x-rays. Analyses of emitted energetic specter allow the accurate determination of the chemical component contained in the sample and also the assessment of its volume.

The result of the realized methods was the quantification of the total content of Fe (iron) wear particles in the assessed gear of the manipulator. At the same time, the run-in processes on the contact surfaces (bearings, gears) were assessed, and they cause the following:

- Change in the micro geometry of surfaces, wear and abrasive (adhesive) wear
- Contamination of the oil filling with solid particles (of abrasive nature), which cause the increase in the wear rate and accelerate the increase in teeth clearance.

By the ferrographic analysis of the lubricating oil sample (analysis of the wear particles), it is possible to identify the real operating condition of the machinery node in terms of quality of the lubrication in rolling or friction contact. In the case of repeated analysis for a certain period of operation, it is possible to determine the evolution of damage or recommend measures to improve the operation in terms of lubrication—friction mode. Ferrographic analysis (volume of 1 ml sample) can detect following wear particles and solid impurities [1, 9]:

1. ADHESIVE FLAKY—arise in the adhesive connection of contacts. The effect of high temperature and pressure leads to the welding of the surfaces, particularly in the case of low bearing capacity or thin oil film.
2. ABRASIVE SPIRAL—formed under hard abrasion. This is often a symptom of operation in emergency mode.
3. FATIGUE, LARGE—long-term load, formation of pitting—fatigue fracture.
4. LAMINAR—flaring of the material, for example, from fatigue fracture.
5. SPHERICAL—typical for the wear of rolling bearing, the occurrence is caused by formation of fine cracks on the contact surfaces.
6. GAMA—HEMATITE—black oxidation, Fe_3O_4, the particles are a symptom of inadequate supply of lubricant to the contact (e.g., lack of lubricant, incomplete oil contact, mechanical or thermal overload...).
7. ALPHA–HEMATIT—red oxidation Fe_2O_3, the particles indicate the presence of the water in lubricant, corrosion of friction contacts.
8. CORROSIVE—occur in the case of chemical breakdown of the lubricant.

9. NON-FERROUS, METALLIC—for example, brass, bronze, non-ferrous metals…
10. NON-FERROUS—for example, fibrous or silica nature, impurities from the surroundings of the operated device.

There are several classification systems for the evaluation of the purity of lubricants. The most common is the method of determining the purity of the lubricant according to standards ČSN 656,206 and ISO 4406. The code of oil purity under the standard ISO 4406 consists of two numbers (e.g., 15/12). The first number is assigned by the number of identified solid particles of size greater than 5 μm. The second number is assigned by the number of identified solid particles of size greater than 15 μm. The assignment of the number of particles to the code numbers is listed in Table 4.1.

Following parameters were identified by the evaluation of the lubricant with respect to the content of impurities (Table 4.2):

- CN—total impurities in the lubricant according to STN 656,210
- The amount of carbon in the lubricant (carbonization residue) indicated in the catalogue of lubricants as CCT number. It determines the degree of thermal oxidation for gear, hydraulic, compressor, turbine oils and to gasoline and diesel engines. The limit value ranges from 2.0% to max. 2.5%.
- DL—the number of large wear particles (over 15 μm)—determined using direct reading magnetic analyzer.

Table 4.1 Representation of the number of solid particles in the lubricant by code classification

Number of particles in 1 ml sample		Code number
More than	To a maximum	
80,000	160,000	24
40,000	80,000	23
20,000	40,000	22
10,000	20,000	21
5,000	10,000	20
2,500	5,000	19
1,300	2,500	18
640	1,300	17
320	640	16
160	320	15
8O	160	14
40	8O	13
20	40	12
10	20	11
5	10	10

Table 4.2 Evaluation of the lubricant condition

Parameter	Condition evaluation			
	Good	Satisfactory	Unsatisfactory	Emergency
CN Total impurities (CCT IR)	–	1.7–2.0%	2.0–2.5%	Over 2.5%
WPC Amount of wear particles	–	To 50%	50–70%	Over 70%
PLP Proportion of macro particles on WPC	–	To 15%	15–35%	Over 35%
H_2O Water content in oil	–	To 2% of the volume	Over 2% of the volume	–

- DS—the number of small and medium wear particles (1—15 μm)—determined using direct reading magnetic analyzer.
- WPC—the amount of wear particles—indicates the level of the system wear.
- PLP—the proportion of macro particles on WPC—indicates the intensity of the lubrication wear.
- Inspection of appearance and color—according to STN 656,076 and ISO 2049.
- Flash point in a closed crucible—according to STN 656,064 and ISO 2719.
- Water content—according to STN 656,062 and DIN 51,777.

The results of the analyses of the condition of the monitored lubricant within the realized experiment were verified by two methods (Table 4.3, Fig. 4.2).

1. Induction, magnetic method using the measuring system Cosmos—Steel Dust Checker
2. X-ray, XRF (x-ray fluorescence method, analysis by x-ray)

The objects of the performed analysis were selected industrial gear oils based on PAO and PG—high performance synthetic oils (HPL class). The analysis of the Fe content (ppm) was performed in gear oil—Optigear RO-220 (samples A—E).

The results of the analysis are summarized in the table below:

The measured values (total content of contaminants, water, small and large particles and Fe) for the sample A are listed in Table 4.4; photos of ferrographics record for objective sample are in Fig. 4.3.

Table 4.3 Results of the realized experiments

Serial number	Sample	LR number	Fe content (ppm)	
			Cosmos	XRF
1	A	4,193	36	800
2	B	4,025	18	760
3	C	7,291	24	840
4	D	7,773	19	1,510
5	E	8,148	38	1,380

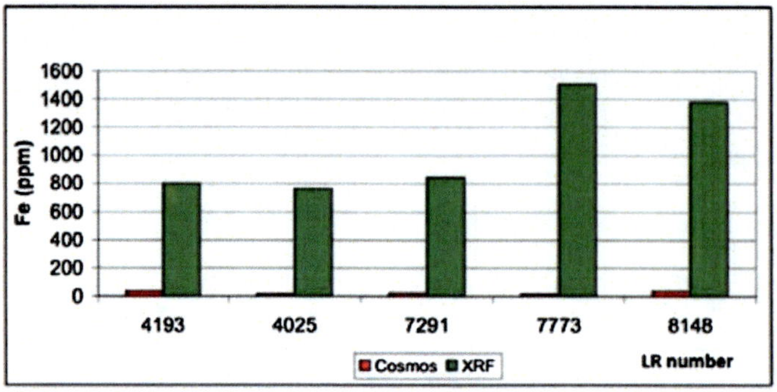

Fig. 4.2 Content of Fe in the gear oil for samples A–E

Table 4.4 The measurement results for sample A

	Total impurities	Water	Large particles	Small particles			Fe content
	CCT IR	H2O	DL	DS	WPC	PLP	(Cosmos)
	[%]	[%]	Number	Number	[%]	[%]	[ppm]
Range	0–2	0–2	0–900	0–900	0–100	0–50	0–1,500
Reality	1.9		40	10	3	1	36

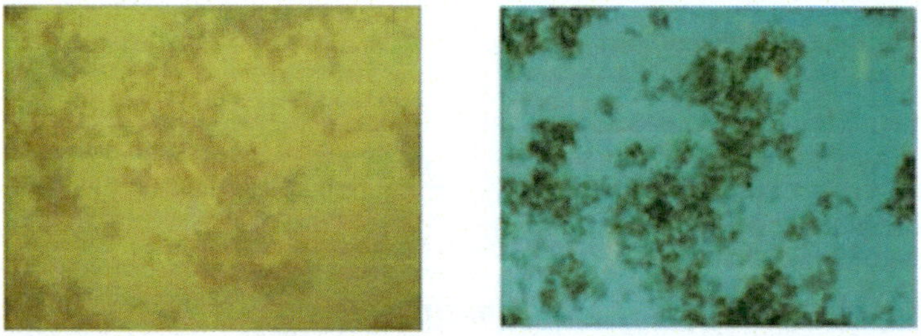

Fig. 4.3 Photos of ferrographic record for sample A—**a** bottom light and **b** upper light

The photos (Figs. 4.3–4.7) show examples of ferrograms—ferrographic traces—obtained for different oil samples, with different content of Fe wear particles.

The measured values (total content of contaminants, water, small and large particles and Fe) for the sample B are listed in Table 4.5; photos of ferrographics record for objective sample are in Fig. 4.4:

The measured values (total content of contaminants, water, small and large particles and Fe) for the sample C are listed in Table 4.6, photos of ferrographics record for objective sample are in Fig. 4.5.

Table 4.5 The measurement results for sample B

	Total impurities	Water	Large particles	Small particles			Fe content
	CCT IR	H2O	DL	DS	WPC	PLP	(Cosmos)
	[%]	[%]	Number	Number	[%]	[%]	[ppm]
Range	0–2	0–2	0–900	0–900	0–100	0–50	0–1,500
Reality	1.9		25	10	2	1	18

Fig. 4.4 Photos of ferrographic record for sample B—**a** bottom light and **b** upper light

Table 4.6 The measurement results for sample C

	Total impurities	Water	Large particles	Small particles			Fe content
	CCT IR	H2O	DL	DS	WPC	PLP	(Cosmos)
	[%]	[%]	Number	Number	[%]	[%]	[ppm]
Range	0–2	0–2	0–900	0–900	0–100	0–50	0–1,500
Reality	2.0		30	15	3	1	24

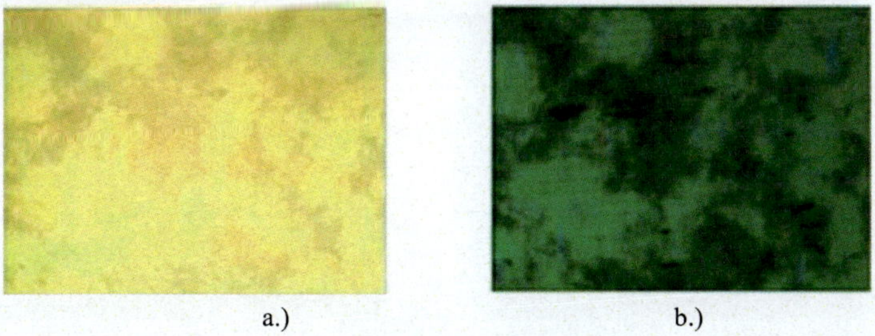

Fig. 4.5 Photos of ferrographic record for sample C—**a** bottom light and **b** upper light

The measured values (total content of contaminants, water, small and large particles and Fe) for the sample D are listed in Table 4.7; photos of ferrographics record for objective sample are in Fig. 4.6.

The measured values (total content of contaminants, water, small and large particles and Fe) for the sample E are listed in Table 4.8, photos of ferrographics record for objective sample are in Fig. 4.7.

Table 4.7 The measurement results for sample D

	Total impurities	Water	Large particles	Small particles			Fe content
	CCT IR	H2O	DL	DS	WPC	PLP	(Cosmos)
	[%]	[%]	Number	Number	[%]	[%]	[ppm]
Range	0–2	0–2	0–900	0–900	0–100	0–50	0–1,500
Reality	2.0		40	15	3	1	19

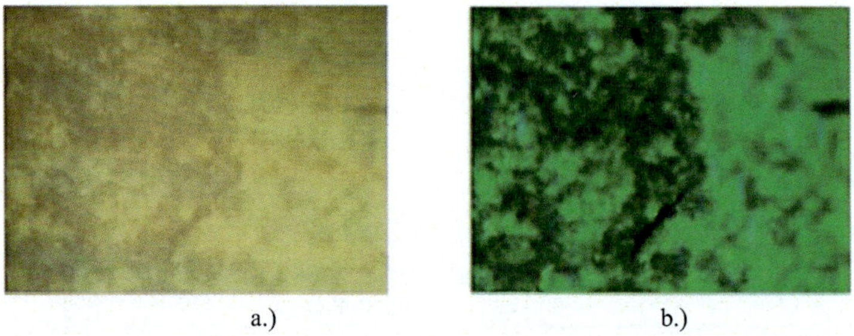

Fig. 4.6 Photos of ferrographic record for sample D—**a** bottom light and **b** upper light

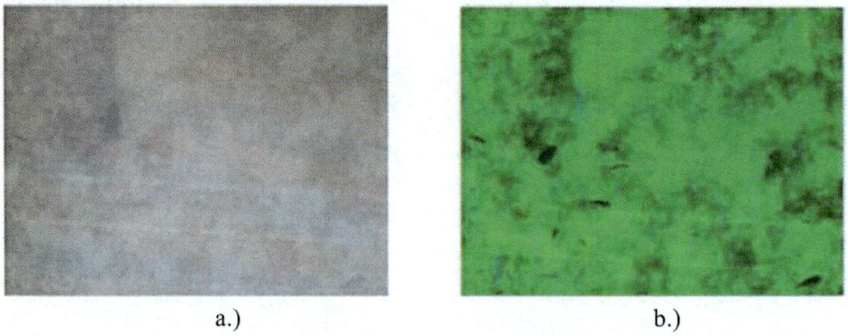

Fig. 4.7 Photos of ferrographic record for sample E—**a** bottom light and **b** upper light

Table 4.8 The measurement results for sample E

	Total impurities	Water	Large particles	Small particles			Fe content
	CCT IR	H2O	DL	DS	WPC	PLP	(Cosmos)
	[%]	[%]	Number	Number	[%]	[%]	[ppm]
Range	0–2	0–2	0–900	0–900	0–100	0–50	0–1,500
Reality	2.0		18	10	2	1	38

4.2.1 Correlation, Quantification of Measured Parameters, Recommended Limits

The realized measurements and the assessment of interrelations show correlation between:

- Individual oil samples, showing the presence and intensity of run-in processes,
- Micro geometrical quality of contact surfaces and surfaces run-ins,
- Increase in kinematic clearance depending on the content of Fe abrasive (adhesive) particles in oil filling (Fig. 4.8).

Completed analyses proved the existence of certain limits for the assessment of the content of Fe particles in gear oils using a simpler method—induction, magnetic method using the measurement system Cosmos—Steel Dust Checker before more sophisticated x-ray fluorescence method, analysis by x-ray (Fig. 4.9).

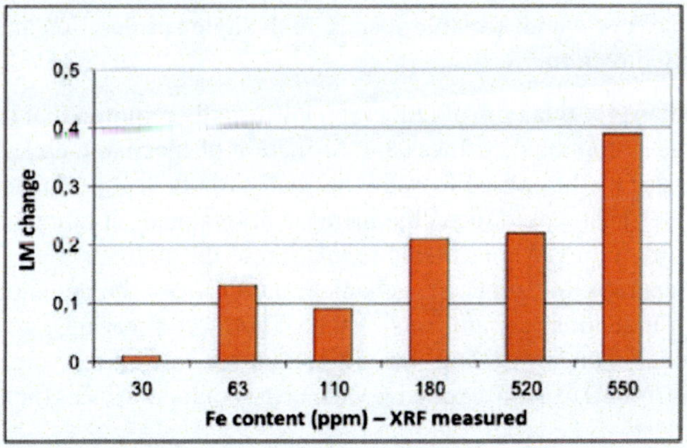

Fig. 4.8 Correlations between LM and Fe content (ppm)

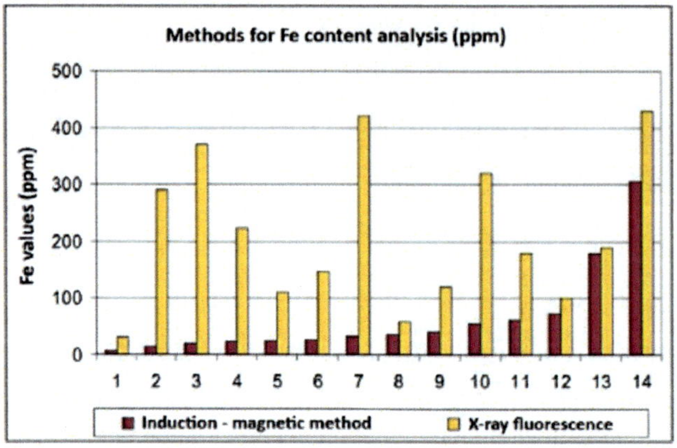

Fig. 4.9 Comparison of the results of analyses by Cosmos and XRF methods

4.2.2 Discussion of Realized Experiments

Cycloid gearboxes used in manipulators and robotic applications are characterized by following operational conditions:

- Low weight and small dimensions,
- High transmitted power (input operating speed and torque),
- High gear ratio (50 to 130),
- High operational temperature of the oil (above 90 °C) and at the same time small volume of oil filling (0.3 to 1.0 kg),
- Operational cycle with variable speeds, high angle acceleration and often change of running direction.

As a standard for this type of reductors, it is usually required that it will not come to the change of torsional stiffness and formation of mechanic clearance (so called lost motion) during the entire life cycle (normally 16.000 h of permanent operation in working cycle. That would mean the increase in torsional vibrations on the gearbox output and difficulties in control and regulation of the actuator.

Cycloid gearbox transmits a mechanical output from the actuator to the output arm through individual components—contact surfaces (bearings, gearings, and the like). For reliable operation, it is very important for entire life cycle of the gearbox (manipulator/robot) to avoid excessive wear at the contact surfaces (adhesive/abrasive friction particles of Fe type).

For consideration of the quality/quantity of emerging Fe friction particles, there is a series of analytical methods for the analyses of oil sample. With use of such methods, we judge the changes of chemical and physical attributes of utilized gear oils. With use of trend analyses of selected parameters of manipulator or robot, we can determine:

- Operational state of gear lubricating oil and the state of contact surface of individual components for transmission of mechanical power,
- In case of increased volume of Fe friction parts, we can understand the tribotechnical mechanism, in which the particles are generated.

Analyses and the knowledge obtained from the test serve as the base for evaluation of the durability of the cycloid gearbox and also as uncommonly valid information for research and optimization of the inner construction of the components for the transmission of the mechanical power of contact surfaces [1, 10, 11].

In the frame scope of realized experiments, we evaluated the state of oil fillings for 5 identical gearboxes (samples labeled as A, B, C, D, and E). We monitored the processes of trial runs of cycloid gearbox form the viewpoint of identification of abrasive particles in first 600 h of operation for individual gearboxes. On the base of the requirement of robotic manipulator producer, primarily two different diagnostic methods were tested—induction method (diagnostic device Cosmos—Steel Dust Checker) and x-ray fluorescent method, for identification of abrasive friction particles on Fe base. Method of ferrography was used for the verification and qualitative description of the friction particles (their type and size).

Experiment pointed out the fact that trial run for observed type of gearing includes the generation of friction particles from contact surfaces of gearings and bearings, which often have the form of magnetic compounds on the Fe base (i.e., oxides such as FeO, Fe_2O_3, Fe_3O_4, and complex compounds, i.e., with additives such as Mo, S, Fe and Ca). The reason for generation of these compounds lies in constructional solution of the cycloid gearbox, which is characterized by small dimensions, what leads to the higher operation temperatures of the oil filling together with an increase in the pressure inside the reductor. There were also particles with dimensions less than 5 μm generated.

Listed circumstances results in notable differences in measured values of Fe volume (ppm) obtained by Cosmos and XRF methods (Table 4.3). While after using of induction method, the values are below the level of acceptable limits (100 Fe ppm), XRF method locates the samples into the area far beyond the recommended limit.

Induction method represents a cheap and quick method for operative, trendy measurement and analyses of the oil sample. Yet after identification of its restrictions, it finds its application with other types of mechanisms, which are not under specific constructional and operational conditions characteristic for cycloid gearboxes (Fig. 4.9).

XRF–RTG ray fluorescent method is more demanding in time and costs, but its application is inevitable in cases, where the sample includes high level of Fe friction particles of small size, which are chemically bound in magnetic form.

Following findings can be concluded on the basis of realized experiments:

1. Induction and magnetic methods for the detection of the content of Fe particles are not sufficiently sensitive to:
   - Small particles of size under 5 μm
   - Non-magnetic particles (for example some Fe oxides or complex Fe and additive compounds from gear oil—wear particles from contact surfaces etc.)
2. Because of the rapid increase in LM, which is undesirable in this type of driving gear, it is important to monitor the total content of Fe particles, including the small and non-magnetic Fe wear particles.

   The content of Fe particles in oil or lubricant was analyzed by Cosmos and XRF methods. The experiment confirmed these arguments:
   - There is a fundamental difference in the measured values of Fe content according to the method used
   - Induction method in this case does not guarantee the correct quantification of measured Fe particles concentrations (wear particles from contact surfaces)

The results of analyses of individual samples using the XRF method confirmed:

1. All examined samples contain a large number of ferrous wear particles,
2. Examined samples simultaneously show a high thermal and oxidative degradation of contact surfaces, the Fe particles are in the form of oxides and complex compounds (Mo, P, S, Ca additives)

The above findings lead to a clear recommendation of the x-ray and XRF methods (x-ray fluorescence methods, analysis by x-ray) for the verification of condition of oils and lubricants for similar types of technical systems and devices.

References

1. Baron, P. - Kočiško, M. - Dobránsky, J. - Pollák, M. - Cmorej, T.: Research and Correlation of Diagnostic Methods for Assessment of the State of Oil Filling in Cycloid Gearbox - 2015.In: Advances in Materials Science and Engineering. P. 97841–97841. - ISSN 1687-
2. Ushakov, I. A. et al.: Handbook of Reliability Engineering, John Wiley And Sons, New York, 1994
3. Dobránsky, J. - Al Hakim, H. – Čačko, P.: Tribotechnical Diagnostics as an Integral Part of Vehicles Service, Technical Weekly, vol. 57, no. 24, pp. 27, 2009
4. Dobránsky, J. – Kreheľ, R.: The Analysis of Oil Quality and Vibration Measurement Center Lathe SV 18RA, Manufacturing Technology, vol. 14, pp. 33-36, 2010
5. J Dobránsky, J. – Kreheľ, R.: Check Oil Quality in Machine Tools Using Tribotechnical Diagnostics, Tribotechnics, vol. 3, pp. 46-47, 2009
6. Czichos, H.: Handbook of Technical Diagnostics: Fundamentals and Application to Structures and Systems, Springer, Berlin, 2012
7. Al Hakim, H. - Blažovský, J.,: Tribológia a servis strojov. Prešov, 2002. 210 s. ISBN 80–7099–875-X
8. Blaškovič, P. - Balla, J. - Dzimko, M.: Tribológia, 1990. 360 s. ISBN 80–05–00 633–0

9. Ragan, E. - Dobránsky, J. - Baron, P. – Olejár, T.: Materials on Dies for Pressure Die Casting, Metalurgija, vol. 51, no. 1, pp. 117-120, 2012
10. Simkulet, V. - Vojtko, I. - Kočiško, M. – Baron, P.: Microstructures Characteristic Evaluation of Carbon Steel C45 After New Turning Method by Master Block, World Academy of Science, Engineering and Technology, no. 78, pp. 664-666, 2013
11. Novák-Marcinčin, J. - Szentivanyi, P. - Janák, M. – Barna, J.: Diagnostic of Manufacturing Devices Operational States by Smart-Phones Application with Use of Augmented Reality Methods, Applied Mechanics and Materials : Operation and Diagnostics of Machines and Production Systems Operational States 2, . Vol. 616, pp. 3–10, 2014

Chapter 5
Application of Technical Diagnostics Tools in the Reductors Test Operation

The running in of reducers is a little explored area in terms of its effect on changing the wide range of characterizing properties of high-precision reducers. At the same time, it is complicated by the nonexistent uniform methodology for the implementation of their run-in. This study describes, in detail, the research of the influence of the trial run of high-precision reducers on the change of their characterizing parameters with the subsequent determination of the methodology for the identification of their critical parameters and positioning accuracy. The research was carried out on a sample of high-precision reducers during a 48-h run-in with the evaluation of changes in their characterizing parameters. The developed methodology unifies the approach to measuring the static and dynamic properties of high-precision reducers to identify their critical parameters and positioning accuracy. The study also points to the need for the correct implementation of the process for the introduction of the bearing reducer into operation after its incorporation into the relevant equipment with emphasis on improving the monitored critical parameters.

5.1 Determination of Methodology and Research of the Influence of the Trial Run of High-Precision Reducers on the Change of Their Characterizing Properties

Technologically advanced devices place increased demands on precise and reliable positioning in industrial applications, especially in robotics and high-precision machines. These requirements lead to the increased use of high-precision cycloidal and harmonic propulsion systems. The reducer, as part of the actuator drive unit, significantly influences its properties (the properties of the entire propulsion system). Reducers used in robotic joints are the most complex subsystems of robots [1–3].

P. Baron et al., *Application of Troubleshooting Tools in the Monitored Production Processes*, Management and Industrial Engineering,
https://doi.org/10.1007/978-3-031-41428-2_5

Recently, these cycloidal, harmonic, and planetary reducers have been brought to perfection to ensure that the robotic system has the properties required. In addition to the requirements for accuracy and reliability, a high gear ratio, high kinematic accuracy, high tilting rigidity, backlash free operation, high torque capacity, compact size, low weight, and high load capacity are also expected [3, 4].

High-precision gear mechanisms (reducers) are used in combination with actuators and form a mechatronic node. The mechatronic node can be described as a coupling member between the non-power (information) control part, and the power control part of the system. Energy conversion occurs in the power section. The mechatronic node can be understood as a complex system consisting of a servomotor, a reducer, a feedback sensor, and possibly also a brake or a redundant position sensor. Such a device of precision mechanics, in combination with control electronics, enables precise control of the position, speed, simplification of kinematic structures, and increase in dynamics while reducing the demands on the assembly of individual sub nodes [3, 5].

Typical for today is highly automated and flexible production aligned with the principles of Industry 4.0. Progressive automation is a fundamental pillar of the modern production process, which requires accuracy, repeatability, and minimal deviations. These factors are a prerequisite for efficient operation.

Typical for today is highly automated and flexible production aligned with the principles of Industry 4.0. Progressive automation is a fundamental pillar of the modern production process, which requires accuracy, repeatability, and minimal deviations. These factors are a prerequisite for efficient operation. An overview of high-precision reducers for industrial robots that operate in accordance with the principles of Industry 4.0 is given in the paper by the team of authors of Pham A.D. et al. [3, 6]. Many research institutes and development centers work on the continuous development and innovation of high-precision reducers. The new design proposal is addressed by authors of several studies, an example being the paper of Blagojevic M. et al. [3, 7], or the paper by Litvin F.L. et al. [3, 7]. In their study, Mao H. et al. [3, 9] present an innovative cycloidal planetary reducer, utilizing the principle of internal engagement instead of an external one between the cycloidal wheel and the pins of the traditional reduction wheel. Its results prove that the proposed transmission power of the reducer is competitive compared to traditional RV reducers. Prototype testing is performed, parameters in individual phases of product life are analyzed, and stress tests are conducted. In their paper, Sun X. et al. [3, 8] focus on the analysis of a new method of modifying the parabolic profile to improve the transmission accuracy, where the result was a difference of up to 5% between the errors measured and the results calculated, thus proving the correctness of this system. The engagement stiffness of the bevel gears in the transmission of the rotational bevel vector was described in the study by Huang Y. et al. [3, 10], which helps researchers to better understand the influence of various parameters on the analysis of the dynamic characteristics of transmission systems. In their paper, the authors of Pham A.D. et al. focused on the analysis of the efficiency of cycloidal reducers with tolerance achieved by means of an approximate force distribution of an ideal cycloidal reducer [3, 11]. The experimental work was carried out by a team of authors (Terakawa T.

et al.) in order to clarify the optimal input formula for precise rotation using a prototype of a motor de-signed by them [3, 12]. In their publication, a team of authors (Gorla C. et al.) addressed the theoretical model, focusing on its fine tuning with the goal of predicting the behavior of the cycloid unit in operation and improving the procedure of designing this unit [3, 13]. In their paper, Yamada S. et al. [3, 14] proposed a precise method of joint torque control with backlash compensation using information on the code on the side of the load application. The efficient use of encoder information on the side of the load application allows the proposed method to compensate for the rapid backlash effect.

Based on the experimental analyses, the new backlash compensation model is proposed to solve the problems caused by the conventional backlash compensation model. In their study, the team of Shih Y.P. et al. [3, 15] deal with the design of the accurate evaluation of cycloidal devices. The research focuses on the creation of a mathematical model of cycloidal equipment in which the areas of profile errors and cog thickness, spacing and jerking errors are tested. The profile points, measured by the coordinate measuring device, are divided into groups of cogs and then formulated as curves using the B-spline curve technique. Currently, the aim of much research is to significantly support the enhancement of their own development base, which ensures innovation and new product development.

To a large extent, the expectations of the current industry from drive units is that the transmission of force effects be done with higher kinematic accuracy. The main part of such a unit is the reducer, generally referred to as a transmission mechanism, whose main function is the transformation of power quantities. They are designed to reduce the speed of the driving machine to the input speed of the driven machine at a constant transmission.

High-precision reducers are based on several mechanical transmission principles:

- Planetary mechanism—reducers with involute gearing.
- Cycloidal mechanism—reducers with trochoidal gearing.
- Harmonic mechanism—reducers with involute gearing.

The manufacturers listed below belong to the world's leading manufacturers of high-precision reducers used in robotics and equipment requiring high-precision motion, each of which produces the following patented reducers:

- Spinea—Slovak manufacturer of high-precision compact bearing reducers TwinSpin. Their global market share is about 3.3%.
- Nabtesco—Japanese manufacturer of high-precision reducers. Their share of the precision compact reducers market is 47%.
- Sumitomo Cyklo—Japanese manufacturer of precision reducers. Their global market share for precision gearboxes is 11.7%.
- Harmonic Drive—Japanese manufacturer of precision reducers. This company uses harmonic transmission in their gearboxes. Their share of the precision compact reducers market is 37.5%.
- Other manufacturers of precision gearboxes with various transformer members share about 0.5% of the world market [3, 16].

The study delves into the research of the influence of the trial run of high-precision reducers on the change in their characterizing properties. As there is no exact trial run methodology for high-precision reducers, only individual manufacturers' recommendations, the research is focused on the influence of their short-term trial run of 48 h on the change in the important individual parameters.

5.1.1 Parameters Characteristic of High-Precision Reducers

The characteristics of high-precision reducers are backlash, dead zone, hysteresis, angular transmission error, high torque capacity, high torsional and tipping rigidity, high efficiency, long life, and small size and weight. Many authors and research teams from all over the world deal with the issue of researching the characteristic properties of high-precision reducers. The issue of backlash is elaborated in more detail in many publications of authors such as Xiong Y. et al. [3, 17], Margielewicz J. et al. [3, 18], as well as in the study by the authors of Getachew A.A. et al. [3, 19] dedicated to monitoring the effect of backlash on transmission error and time-varying mesh stiffness. The simulation and analysis of the exact operation of robots and gear tooth contact were discussed by the authors Spitas C. et al. [3, 20] and Walha L. et al. [3, 21] in their publications. Angular measurements of high-precision reducers used in robots have been investigated by Xue J. et al. [3, 22]. Many mathematical tools are used to process calculations and model control systems, which were discussed in the publication by Caldarola F. et al. [3, 23]. The nonlinearity of the behavior of transmission systems and their analysis was discussed in more detail by authors Farshidianfar A. et al. [3, 24].

Backlash (Backlash in Teeth)

The term backlash is defined as the backlash between adjacent mechanical parts of the transmission gears, or as the value by which the width of the space transcends the tooth on the pitch circle. The backlash has a negative effect on the precise control of actuators containing a gear mechanism, the properties of which are important parameters with respect to the use of actuators or higher nodes.

A generally acceptable model of static backlash in the teeth represents the backlash in the teeth, where v is the output, u is the input, m represents the incline, and $\alpha > 0$ is half the width of the gap representing the backlash in the teeth. In gearboxes, the width of the gap means the total width or the gap between the coengaging gears of the gearbox. The model of Fig. 5.1 is simplified, since in the actual gearbox the initial position of the teeth does not lie in the middle [3, 25].

The static backlash in the teeth is represented by two parallel lines—one ascending and one descending, connected by a horizontal line, representing the backlash in the teeth itself. The ascending line is active if the input v(t) and the output u(t) grow, i.e.,

$$u(t) = m(v(t) + \alpha), \quad \text{if}: \ m > 0, \dot{v}(t) > 0 \tag{5.1}$$

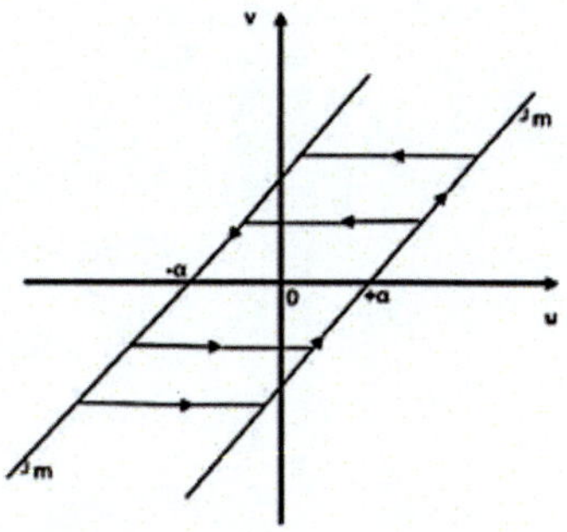

Fig. 5.1 Backlash in the teeth model

The descending line is active if v(t) and the u(t) simultaneously fall, i.e.,

$$u(t) = m(v(t) - \alpha), \quad \text{where}: \ \dot{v}(t) < 0 \tag{5.2}$$

If the tooth of the gear is located in the gap between the two teeth of the engaging wheel, then the following holds: $\dot{u}(t) = 0$.

Mathematical representation of the backlash in the teeth:

$$\dot{u}(t) = \begin{cases} m\dot{v}(t) & \\ & \text{if} \ \ \dot{v}(t) > 0 \ \text{ and } \ \dot{u}(t) = m(v(t) + \alpha) \\ & \text{or} \ \ \dot{v}(t) < 0 \ \text{ and } \ u(t) = m(\dot{v}(t) - \alpha) \\ 0 & \end{cases} \tag{5.3}$$

Hysteresis

The static backlash in the teeth is represented by two parallel lines one ascending and one descending, connected by a horizontal line, representing the backlash in the teeth itself.

Hysteresis is one of the basic features of gearboxes. This characteristic conveys information about the static, as well as dynamic, properties of the examined gearbox. The basic and, at the same time, the most important characteristics that can be obtained from the hysteresis curve on the gearbox subjected to the measurement include torsional rigidity, backlash in the teeth and positioning accuracy. “If the input shaft and the body are secured against rotation and the torque acts on the output flange, the graphical representation of the load has the shape of a hysteresis curve.” Thus, hysteresis is an undesirable factor that contributes to faster wear (Fig. 5.2).

In general, the hysteresis curve is an undesirable factor, adversely affecting positioning accuracy and wear. The hysteresis curve of the gearbox can be considered a quasistatic property of the gearbox, as it is measured; therefore, it is not possible to compensate for this property while the gearbox is running.

The hysteresis curve of the gearbox is a variable nonlinearity, which depends on the position of rotation of the output flange, whose hysteresis curve is to be measured, and on the load moment acting on the reducer. Different changes in the hysteresis

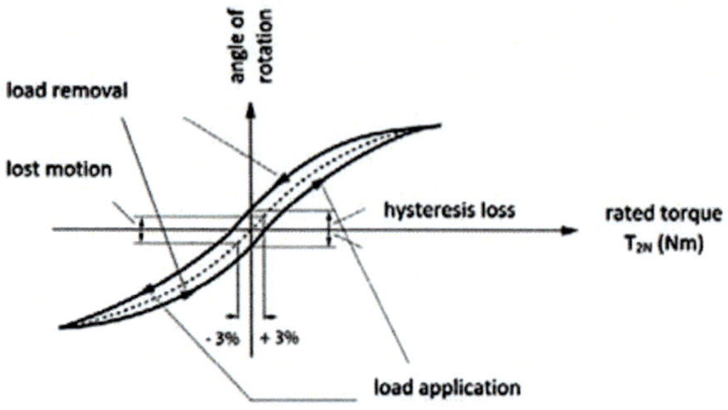

Fig. 5.2 Hysteresis curve typical for Sumitomo Fine Cyclo/Spinea TwinSpin

characteristic of the same gearbox emerge when the hysteresis curve is measured in several positions of the output position. The difference is due to imperfect production and assembly despite the high precision of the production equipment, and also as a result of the pairing system of individual components of which the gear unit is composed. This fact points to the uniqueness of each gearbox, which, despite the same production technology and the effort to approach the ideal parameters of the individual components, has its own unique properties.

Angular Transmission Error

The angular transmission error (ATE) is the sum of the maximum positive and negative differences between the theoretical and actual rotation angle of the rotary module output. It represents the error between the input angle of the shaft rotation and the output angle of the flange rotation of the rotary module. Knowing this characteristic of the rotary module (in various connection designs) is especially important for the uniformity, smoothness, accuracy and repeatability of the output position.

As Xu H. et al. [3, 26] write in their paper, the transmission error is generally measured under the conditions of zero load and low speed. However, low speed is a general concept and there is no general standard of speed measurement or proper scientific grounds for its definition. Therefore, it is difficult to obtain consistent transmission accuracy evaluation results for the same reducer.

ATE is measured at a constant velocity and a very small angular speed without any load to eliminate the hysteresis effect of fast dynamic events. The actual velocity of the output position depends on the input position of the inlet flange and the ATE value in this position [3, 27].

This dependence can be described as follows:

$$\omega_{out_{(\varphi_{in})}} = i_{(\varphi_{in})} \cdot \omega_{in} = \left(i_{th} + \Delta i_{(\varphi_{in})}\right) \cdot \omega_{in} \tag{5.4}$$

By integrating this equation, we get a change in the output position depending on the ATE:

$$\Delta\varphi_{ch_{(\varphi_{in})}} = \int \Delta\omega_{out}\, dt = \int \Delta i_{(\varphi_{in})} \cdot \omega_{in}\, dt = \omega_{in} \int \Delta i_{(\varphi_{in})}\, dt \tag{5.5}$$

ATE Δi can also be expressed using the harmonic function:

$$\Delta i_{(\varphi_{in})} = \sum\nolimits_k \Delta i_k \cdot \sin\left(\frac{2\pi}{T_k}\varphi_{in} + \gamma_k\right), \quad k > 0 \tag{5.6}$$

where T_k is the length of the period on the side of the drive φin is the angle of rotation of the transmission input flange.

By adding Eq. (5.6) into Eq. (5.5), the following derivation $\Delta_{\varphi ch(\varphi in)}$ is obtained:

$$\Delta\varphi_{ch_{(\varphi_{in})}} = \omega_{in} \sum_k \Delta i_k \cdot \int sin\left(\frac{2\pi}{T_k}\varphi_{in} + \gamma_k\right) dt \tag{5.7}$$

where given the constant speed when measuring ATE,

$$\Delta\varphi_{ch_{(\varphi_{in})}} = \omega_{in} \sum\nolimits_k \Delta i_k \cdot \int sin\left(\frac{2\pi}{T_k}\varphi_{in} + \gamma_k\right)$$
$$dt = \sum\nolimits_k \frac{T_k.\Delta i_k}{2\pi} \cos\left(\frac{2\pi}{T_k}\varphi_{in} + \right)\gamma_k \tag{5.8}$$

Dead Zone

The dead zone is measured similarly to ATE $\Delta\varphi_{ch}$. The difference is that during the entire measuring range, the positive, as well as the negative, direction of rotation is measured at the same time and not separately, as is the case with ATE. The dead zone represents the difference of the following deviations in the same position of the output flange rotation:

$$\Delta\varphi_{mch} = \Delta\varphi_{ch}^{-} - \Delta\varphi_{ch}^{+} \tag{5.9}$$

It should be noted that each manufacturer has its own terminology, which differs in the individual descriptions of the same quantities, e.g., backlash, lost motion (LM), and dead zone. Many times, this quantity is also referred to as the gearbox backlash, and such a designation is a bit confusing with respect to precision gearboxes with quasizero backlash. In case of the TwinSpin Spinea gearboxes, the dead zone is called based on the hysteresis characteristic as follows:

The dead zone (DZ) is the angle of rotation of the output flange at ±3% of the nominal torque measured at the center of the hysteresis curve (Fig. 5.3). If the input shaft and the body are secured against rotation and the torque acts on the output flange, DZ can be measured under such conditions 0.

The definition of torsional rigidity is based on this characteristic, defined by the following relation:

$$k_t = \frac{c}{d} \text{ [Nm/arcmin]} \tag{5.10}$$

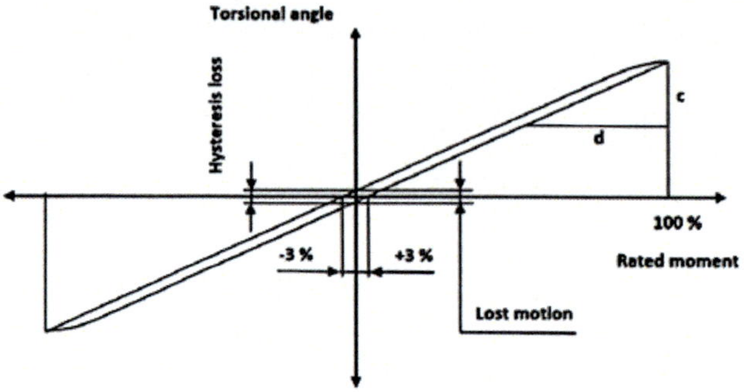

Fig. 5.3 Dead zone definition based on the hysteresis curve

Other Parameters Characteristic of High-Precision Reducers

- Starting torque—a load-free, quasi-static torque is required to start the rotation of the input shaft if no load is applied to the output flange.
- Reverse torque—the torque applied to the output flange that is required to start the rotation of the input shaft, to which no load is applied.
- Tipping rigidity—high-precision gearboxes transmit external forces and torque loads using integrated output bearings. When the load is applied to the output flange, the angle of deflection of the flange is proportional to the respective tilting torque. The torque rigidity Mt is the tipping torque at which the output flange deviates by an angle of $\Theta = 1'$.
- Effective input speed—the limit value of the average speed of the working cycle.
- Maximum continuous input speed—the speed limit value in continuous operating mode, S1.
- Maximum acceleration and braking torque—due to the inertial load, the torque acting during acceleration and braking is higher than the nominal value.

5.1.2 Description of the Investigated Problem

The primary diagnostics of the trial run impact will focus on Spinea, the Slovak manufacturer of cycloidal bearing reducers, TwinSpin. It is necessary to point out the design of the TwinSpin bearing reducer, which integrates roller bearings in the radial and axial arrangement of the rolling elements. By implementing assembled roller bearings, space savings are achieved under a controlled maximum carrying capacity. The TS 050 semihollow bearing reducer (BR) was chosen as the basic representative for diagnostics purposes (Table 5.1).

The measurement was done on a diagnostic device, which allows the measurement to be performed in two required configurations, as follows.

Table 5.1 Conditions of measurements applied to TS 050

TS 050		
Gear ratio	i	63
Nominal output torque 100%	TR 100% [Nm]	18
Nominal output torque 50%	TR 50% [Nm]	9
Rated output speed	nR [rpm]	1000
Vibration measurement	–	Yes
Lubricant analysis	–	Yes
Amount of lubricant	[cm^3]	2.85
Lubricant type		Castrol Tribol TT1 PD

5.1.2.1 Measurement of BR Characteristics—Configuration A

The measured bearing reducer (BR) is driven by a servomotor from the side of the BR input flange and a mechanical weight (load) is applied onto its arm on the BR side of the output flange (Fig. 5.4). A bellows coupling was used to compensate for the angular misalignments, radial misalignments and axial misalignments of the connected shafts, as well as to ensure accurate transmission of angular motion and torque. Figure 5.5 shows a realistic view of the measuring stand.

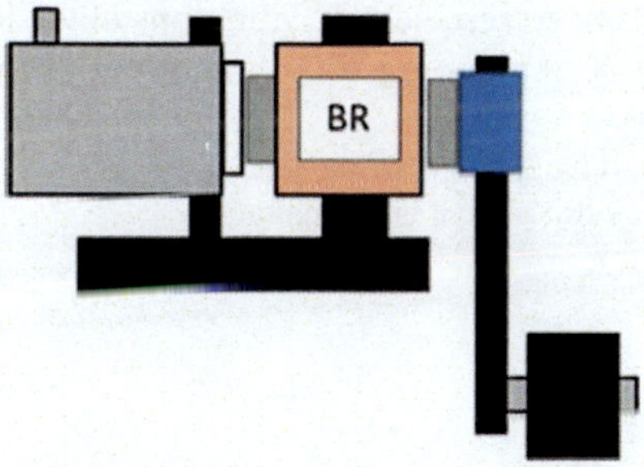

Fig. 5.4 BR connection arrangement in configuration A

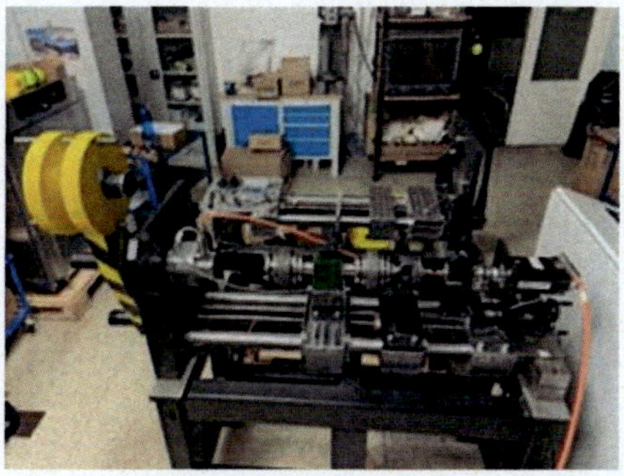

Fig. 5.5 View of the stand in configuration A

5.1.2.2 Measurement of BR Characteristics—Configuration B

The bearing reducer (BR) subjected to measurement is located on a common axis with the driven and loading servomotor (Fig. 5.6). The loading drive forms a connection between the planetary reduction mechanism and the loading servomotor. In this configuration, the measuring set includes an external position sensor and an external torque sensor. A bellows coupling was used to compensate for the angular misalignments, radial misalignments and axial misalignments of the connected shafts, as well as to ensure accurate transmission of angular motion and torque. Figure 5.7 shows a realistic view of the measuring stand.

The following analyses were performed to assess the change in the characteristics of the bearing reducers during their trial run:

- Lubricant analysis,
  - changes in surface microgeometry, wear or abrasive (adhesive) abrasion. Contamination of the oil fill with solid particles of an abrasive nature, which cause an increase in the rate of wear and accelerate the increase in the BR backlash;
- Analysis of temperature changes;
- Analysis of changes in starting torque—fluctuations;
- Analysis of H, LM changes;
- Analysis of changes in the system rigidity;
- Assessment of vibrations in the low frequency range;
- Assessment of vibrations in the high frequency range.

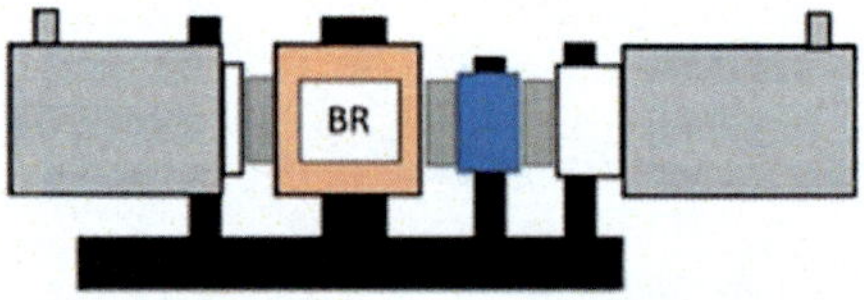

Fig. 5.6 BR connection arrangement in configuration B

Fig. 5.7 View of the stand in configuration B

5.1.3 Characteristics of the Diagnostic Methods Applied

Lubricating oil analyses:

- ppm Fe abrasive particles were evaluated.

The following were applied to assess vibrations in the low frequency range:

- Velocity measurement method (jitter speed), [mm/s] unit, RMS detection.
- According to the ISO 10816–3: 2009 recommendation, FFT (fast Fourier transform) spectrum for the frequency range 10–800 Hz.
- Enveloping Acceleration measurement method (method of measuring vibration acceleration in the high frequency range up to 10–20 kHz, through which the condition of roller bearings and gears is assessed), gE unit, PtP detection, for the frequency range 50—1000 Hz, FFT spectrum, and time log.

The following were used to assess vibrations in the high frequency range:

- (Jitter) acceleration measurement method, g unit, PtP detection, frequency range up to 16 kHz, FFT spectrum, and time log
- Enveloping acceleration measurement method, gE unit, PtP detection, frequency range up to 10 kHz and up to 20 kHz, FFT spectrum, and time log.
- Diagnostic assessment methods in the field of acoustic emission and ultrasound:
- SEE (spectral emitted energy) measuring method—acoustic emissions (SKF), See unit, PtP detection, frequency range up to 600 kHz, FFT spectrum, and time log.
- HFD measuring method—high frequency detection method (SKF), analysis of oscillations (accelerations) in the frequency range 40 kHz.
- Measuring method is used to assess lubrication, carrying capacity of oil film, metal contact, seizure, abrasion of contact surfaces (bearings and gears).

The so-called bump test was done on the BR—the search for natural frequencies of the measured system in the torsional and tipping direction. This is a traditional method of measurement. Performing a bump test involves the use of at least a two-channel frequency analyzer, impact hammer, and accelerometer.

During the measurement, the LR TS 050 was fixed to the diagnostic equipment frame with screws. A mechanical arm with a weight of 2.5 kg was attached to its outlet flange. The excitation force was generated by the excitation impact hammer at the location of the weight at the end of the weight arm.

During the bump test, the oscillation amplitudes and frequencies are monitored by accelerometers and displayed as the area of the amplitude—frequency dependence. The most important information obtained is in the frequency deviations of the LR structure at their locations, and to a lesser extent, the attenuation values associated with these modes.

Subsequently, the BR was measured at a constant rotational frequency of 1000 rpm at the input at a sinusoidal load (weight on the arm) in both directions of rotation (CW, CCW). The vibration sensor, PBC 356B18, was placed on the support body of the gearbox (in the vertical direction). At the same time, a repeated measurement of the startup ramp was performed—the measurement of absolute vibrations during a continuous change of speed at the input, in the range of 0–2000 rpm and 0–3000 rpm. The aim of this measurement was to compare the effect of the onset of resonant events–amplification of the output arm vibration. Vibration sensors were placed on the output arm, at the point of the weight, in the radial and axial direction.

Temperature measurement was done on the BR in order to monitor temperature changes (BR heating) during the trial run process.

5.1.4 Conducting Measurements of Characteristic Properties of Bearing Reducers During Their Trial Run

5.1.4.1 Bearing Reducer Trial Run—Configuration A

The measurement was performed in a type A configuration, i.e., the load was applied onto the arm by means of a mechanical weight. The measurements of the examined object were performed on 10 samples with a load of 50% TR at a constant rotational input frequency of 1000 rpm over 48 h. During the trial run mode, the temperature on the surface of the BR support body was measured and logged. Abrasion particle measurements were taken at eight—hour intervals throughout the BR trial run period.

Bearing reducer TS 050 parameters measured before the trial run in configuration A are shown in Table 5.2.

A measurement of the starting BR TS 050 torque values was performed under the following parameters:

- Input rotation speed: 10 rpm.
- Number of output flange revolutions completed: 1 revolution.

Table 5.2 Bearing reducer TS 050 parameters measured before the trial run (configuration A)

No	Type BR/serial No	Lost motion [arcmin]	Hysteresis [arcmin]	Rm (CW) [Nm]	Rm (CCW) [Nm]
1	TS 050 s/n.1901	0.57	0.42	0.030	0.036
2	TS 050 s/n.1902	0.64	0.55	0.031	0.023
3	TS 050 s/n.1903	0.57	0.44	0.027	0.021
4	TS 050 s/n.1904	0.52	0.55	0.024	0.017
5	TS 050 s/n.1905	0.53	0.45	0.027	0.033
6	TS 050 s/n.1906	0.55	0.42	0.032	0.024
7	TS 050 s/n.1907	0.69	0.41	0.039	0.044
8	TS 050 s/n.1908	0.55	0.53	0.030	0.027
9	TS 050 s/n.1909	0.51	0.52	0.039	0.044
10	TS 050 s/n.1910	0.64	0.44	0.037	0.043

- Number of shaft revolutions completed: 63 revolutions.
- Sensor sampling frequency: 200 smps/s.
- Sensor resolution: 1:1.
- Ambient temperature: 21 °C ± 2.

Test cycle:

- Torsional moment Mt (m = 4.0 kg on the arm of 0.425 m).
- When measuring Mt loaded to tipping moment Mc = 44 Nm.
- Average cycle torque Tave = 16 Nm.
- Average input speed nave = 2054 rpm.
- No cycle pause ≥ load factor Z = 100%.
- Acceleration time to 1000 rpm tacc = 0.5 s.
- Reverse cycle S8.

The measurement of the Fe solids' concentration was carried out during the entire BR trial run in the (mechanical) weight application configuration, according to the conditions set out in Table 5.3. The measurement of temperature changes on the surface of the BR support body was performed using a sensor with a built-in magnet of the Alnico type, which is suitable for measuring the surface temperature of steel surfaces. The sensor was attached with a magnet. Temperature measurements were performed during the entire period (48 h) of the trial run at 1-h intervals. During the measurement, the alarm value was set at 60 °C, which is the maximum permissible value set by the manufacturer on the BR support body. The summary values of the surface temperatures BR are shown in Fig. 5.8.

After 48 h of the continuous trial run mode, diagnostics were performed by means of the so-called bump test. The aim of the test was to determine the natural frequencies of the measured BR 050 system in the torsional and tipping direction. The vibration sensor of the triaxial accelerometer PBC 356B18 was glued to the support body in the vertical direction. The accelerometer was mounted on a mechanical arm. The summary of the natural frequency values measured in the axial and radial direction for the individual BR samples is shown in Table 5.4.

Based on the conducted measurements, we can state that the natural frequency in the radial direction ranges between 820 ± 10 rpm. At the same time, it can be stated that the natural frequency (critical speed) in the axial direction ranges between 1260 ± 50 rpm. The starting torque (Rm) BR TS 050–63, s/n.1901 before the trial run of CW direction is shown in Fig. 3.7, and that for CWW direction is in (Figs. 5.9 and 5.10).

Following the trial run mode and the collecting of the measurements of the reducer's natural frequencies, a summary of the initial information about the bearing reducers TS 050 with a gear ratio of 63, necessary for evaluating the parameter changes was obtained. Subsequent to the trial run, measurements of the LM, H, and Rm (CW—Fig. 5.11, CCW—Fig. 5.12) parameters were conducted again, as they could have been affected by the trial run mode. The summary of the LM, H, and Rm (CW, CCW) values measured for individual samples of the monitored BR parameters

Table 5.3 Values of solid abrasive particles during 48 h of the BR TS 050 trial run

No	Type BR/ serial No	Content Fe (ppm) after 8 h	Content Fe (ppm) after 16 h	Content Fe (ppm) after 24 h	Content Fe (ppm) after 32 h	Content Fe (ppm) after 48 h
1	TS 050 s/ n.1901	7	10	12	14	16
2	TS 050 s/ n.1902	8	12	13	15	16
3	TS 050 s/ n.1903	9	11	14	16	17
4	TS 050 s/ n.1904	10	13	16	18	18
5	TS 050 s/ n.1905	9	12	15	17	18
6	TS 050 s/ n.1906	11	12	14	16	17
7	TS 050 s/ n.1907	8	11	13	15	16
8	TS 050 s/ n.1908	12	13	16	18	19
9	TS 050 s/ n.1909	9	14	16	17	17
10	TS 050 s/ n.1910	10	13	14	16	17

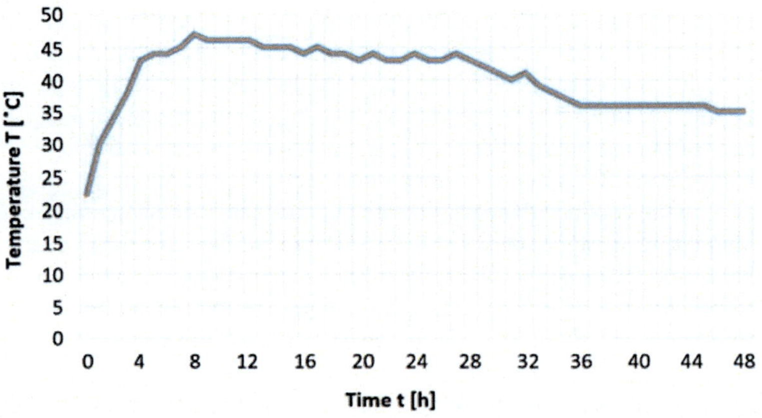

Fig. 5.8 Graphical display of the courses of the temperature on the surface of the BR TS 050 support body during the trial run

Table 5.4 Values of natural BR TS 050 frequencies measured after the trial run (configuration A)

No	Type BR/ serial No	Natural frequency radial—torsional [RPM/Hz]	Natural frequency axially—tipping [RPM/Hz]
1	TS 050 s/ n.1901	840/14.0	1260/21.0
2	TS 050 s/ n.1902	825/13.0	1210/20.0
3	TS 050 s/ n.1903	831/12.5	1245/21.0
4	TS 050 s/ n.1904	844/14.0	1270/20.0
5	TS 050 s/ n.1905	837/13.0	1233/19.0
6	TS 050 s/ n.1906	833/12.0	1250/18.5
7	TS 050 s/ n.1907	842/14.0	1280/21.0
8	TS 050 s/ n.1908	830/13.0	1310/22.0
9	TS 050 s/ n.1909	820/11.0	1240/19.0
10	TS 050 s/ n.1910	835/11.5	1260/20.0

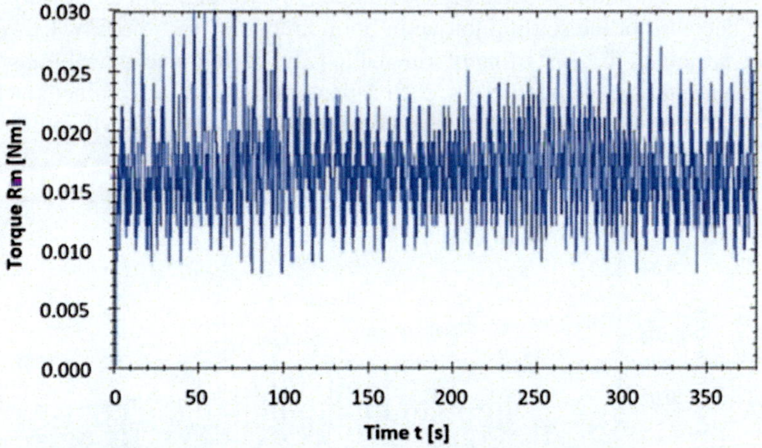

Fig. 5.9 Starting torque (Rm) BR TS 050–63, s/n.1901—CW direction (before the trial run)

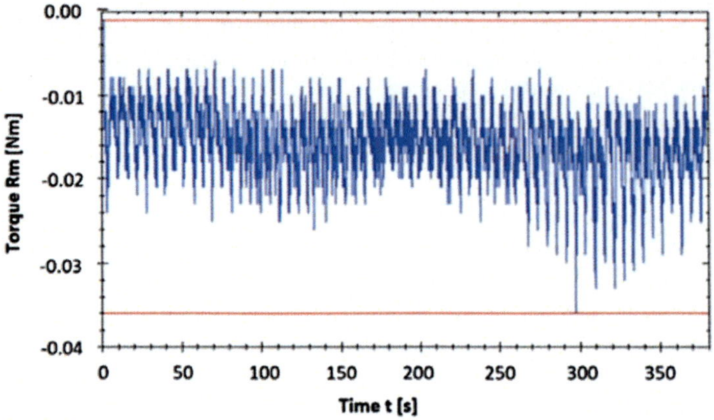

Fig. 5.10 Starting torque (Rm) BR TS 050–63, s/n.1901—CCW direction (before the trial run)

is shown in Table 5.5, and the course of the hysteresis curve BR TS 050–63, s/n.1901 after a 48 h-long trial run, in Fig. 5.13.

Summary of partial results from the application of MDS during the trial run of the TS 050 bearing reducer—weight load (mechanical weight at the end of the arm) applied during the trial run.

Measurements were conducted on 10 BR 050–63 samples. Before the trial run, we performed a series of measurements on the samples to identify the factual state of BR. After an initial 48 h trial run with a load factor of 9 Nm at the output, there was a decrease in the starting torque in both directions, i.e., the CW, CCW, which can be attributed to the starting of kinematic pairs during mechanical transmission. After the trial run, the parameters, such as lost motion (LM), remained unchanged. In the case of the hysteresis parameter (H), there was a slight growth in the increase

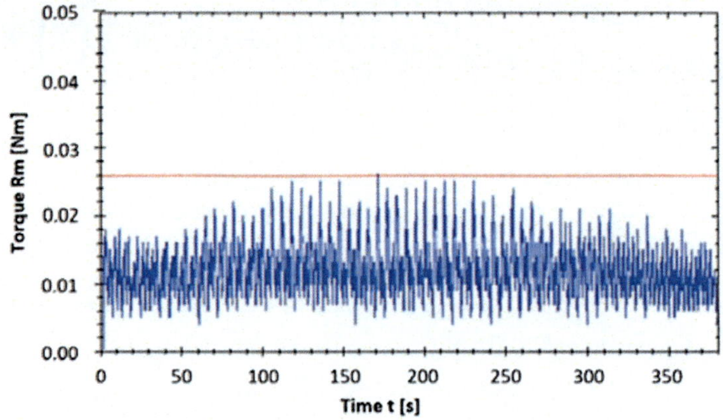

Fig. 5.11 Starting torque BR TS 050–63, s/n.1901—CW direction (after the trial run)

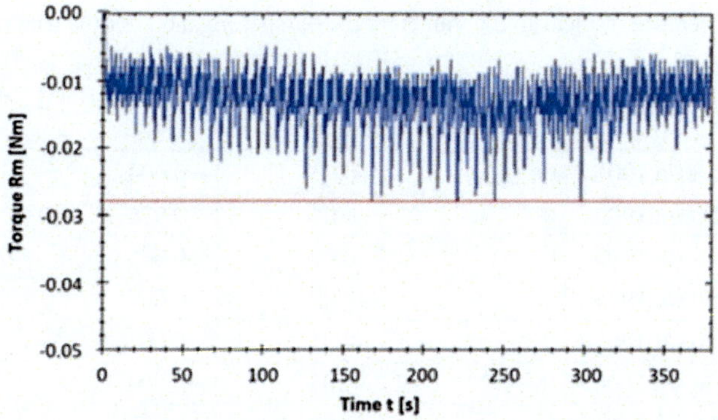

Fig. 5.12 Starting torque BS TS 050–63, s/n.1901—CCW direction (after the trial run)

Table 5.5 Bearing reducer TS 050 parameters measured after the trial run (A)

No	Type BR/serial No	Lost motion [arcmin]	Hysteresis [arcmin]	Rm (CW) [Nm]	Rm (CCW) [Nm]
1	TS 050 s/n.1901	0.57	0.45	0.026	0.023
2	TS 050 s/n.1902	0.64	0.59	0.025	0.019
3	TS 050 s/n.1903	0.57	0.51	0.021	0.017
4	TS 050 s/n.1904	0.52	0.64	0.019	0.015
5	TS 050 s/n.1905	0.53	0.49	0.022	0.024
6	TS 050 s/n.1906	0.55	0.47	0.028	0.020
7	TS 050 s/n.1907	0.69	0.45	0.032	0.032
8	TS 050 s/n.1908	0.55	0.62	0.024	0.023
9	TS 050 s/n.1909	0.51	0.59	0.030	0.038
10	TS 050 s/n.1910	0.64	0.52	0.029	0.037

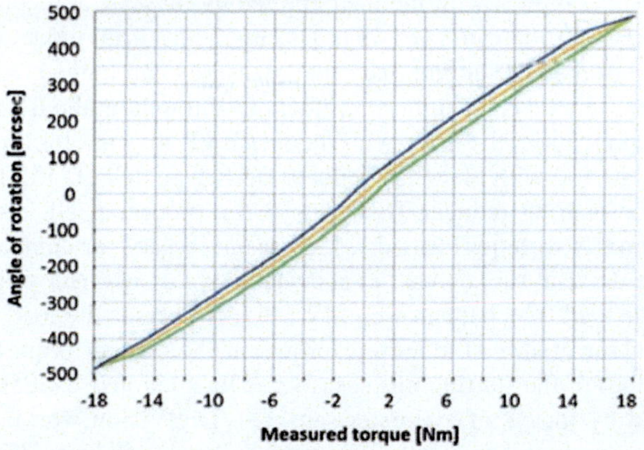

Fig. 5.13 Course of hysteresis curve BR TS 050–63, s/n.1901 after 48 h-long trial run

Table 5.6 Difference of measured parameters LR 050 before and after the trial run (configuration A)

No	Type BR/serial No	Lost motion [Arcmin]	Hysteresis [Arcmin]	Rm (CW) [Nm]	Rm (CCW) [Nm]
1	TS 050 s/n.1901	0	+0.03	−0.004	−0.013
2	TS 050 s/n.1902	0	+0.04	−0.006	−0.004
3	TS 050 s/n.1903	0	+0.07	−0.006	−0.004
4	TS 050 s/n.1904	0	+0.09	−0.005	−0.002
5	TS 050 s/n.1905	0	+0.04	−0.005	−0.009
6	TS 050 s/n.1906	0	+0.05	−0.004	−0.004
7	TS 050 s/n.1907	0	+0.04	−0.007	−0.012
8	TS 050 s/n.1908	0	+0.09	−0.006	−0.004
9	TS 050 s/n.1909	0	+0.07	−0.009	−0.004
10	TS 050 s/n.1910	0	+0.08	−0.008	−0.006

in value, but the value of the difference itself is so small that no significant effect needs to be attributed thereto. The range of change is 0.03 to 0.09 arcmin (Table 5.6).

From the course of the temperature values measured on the support body after the initial temperature rise to 47 °C, it is apparent that the temperatures stabilized at 36 °C after about 35 h. The maximum degree of contamination of the PG TT1 lubricant with Fe solid particles was 19 ppm Fe.

5.1.4.2 Trial Run of the Gear Reducer in Configuration B

The measurement was conducted in a type B configuration, i.e., under the load from the load applying servomotor. The measurements of the examined object No. 1 were conducted on 10 samples with the load applied to 50% TR at a constant rotational input frequency of 1000 rpm over 48 h. During the trial run mode, the temperature on the surface of the BR support body was measured and logged. Abrasion particle measurements were taken at eight-hour intervals throughout the BR trial run period. As the load application, the drive with reducer B&R servomotor type 8LSA55 was used (DA030S000-3 with planetary reducer B&R type 8GP50-155-050S1L5).

Prior to the bearing reducer TS 050 trial run process itself, the initial parameter measurements were performed, which are necessary for comparing diagnostic changes after the trial run, as well as for qualitative assessment of the influence of the type of the load (mechanical weight on the arm versus the load applying drive) during the trial run mode on the starting torque and the starting torque fluctuation. The measured values of the starting torques (torque fluctuations) BR 050 were recorded and evaluated by means of a Burster sensor + DigiVision working environment, similar to the measurements in configuration A. Bearing reducer TS 050 parameters measured before the trial run in configuration B are shown in Table 5.7.

Table 5.7 Bearing reducer TS 050 parameters measured before the trial run (configuration B)

No	Type BR/serial No	Lost motion [arcmin]	Hysteresis [arcmin]	Rm (CW) [Nm]	Rm (CCW) [Nm]
1	TS 050 s/n.1911	0.73	0.55	0.045	0.043
2	TS 050 s/n.1912	0.65	0.95	0.035	0.041
3	TS 050 s/n.1913	0.80	1.11	0.040	0.046
4	TS 050 s/n.1914	0.78	0.63	0.032	0.038
5	TS 050 s/n.1915	0.62	1.00	0.042	0.050
6	TS 050 s/n.1916	0.71	0.82	0.051	0.046
7	TS 050 s/n.1917	0.82	1.04	0.036	0.040
8	TS 050 s/n.1918	0.90	0.74	0.044	0.042
9	TS 050 s/n.1919	0.55	0.67	0.032	0.039
10	TS 050 s/n.1920	0.76	0.90	0.028	0.034

Measurement of the starting BR TS 050 torque values was performed under the following parameters:

- Input rotation speed: 10 rpm.
- Number of output flange revolutions completed: 1 revolution.
- Number of shaft revolutions completed: 63 revolutions.
- Sensor sampling frequency: 200 smps/s.
- Sensor resolution: 1:1.
- Ambient temperature: 21 °C ± 2.

Test cycle:

- Torsional moment caused by the load applying drive.
- Average cycle torque Tave = 16.0 Nm.
- Average input speed nave = 2054 rpm.
- Acceleration time to 1000 rpm tacc = 0.5 s/1000 rpm.
- Reverse cycle S8.

The measurement of the Fe solids' concentration (Table 5.8) was performed during the entire BR trial run period in the load applied by drive configuration (motor versus motor). The measurement was conducted with an SDM-72 evaluation device. The graphic display of the course of temperature changes during the trial run is shown in Fig. 5.14.

After a 48-h continuous trial run mode, diagnostics were performed to determine the natural frequencies of the measured BR 050 system in the torsional direction, only. Since only a torsional load can be generated by the load applying servomotor, the natural frequencies in the tipping direction have not been identified. The vibration sensor of the triaxial PBC 356B18 accelerometer was glued to the support body in the vertical direction. The accelerometer was mounted on the mechanical bracket of the frame. The summary of the natural frequency values measured in the axial

Table 5.8 Solid abrasive particle values over 48 h of the BR TS 050 trial run

No	Type BR/ serial No	Content Fe (ppm) after 8 h	Content Fe (ppm) after 16 h	Content Fe (ppm) after 24 h	Content Fe (ppm) after 32 h	Content Fe (ppm) after 48 h
1	TS 050 s/ n.1911	9	10	10	14	14
2	TS 050 s/ n.1912	7	8	11	12	12
3	TS 050 s/ n.1913	11	12	12	14	15
4	TS 050 s/ n.1914	8	10	11	12	13
5	TS 050 s/ n.1915	7	9	10	11	13
6	TS 050 s/ n.1916	9	11	12	14	14
7	TS 050 s/ n.1917	8	9	11	13	13
8	TS 050 s/ n.1918	7	10	12	14	14
9	TS 050 s/ n.1919	9	9	11	13	14
10	TS 050 s/ n.1920	7	8	9	11	11

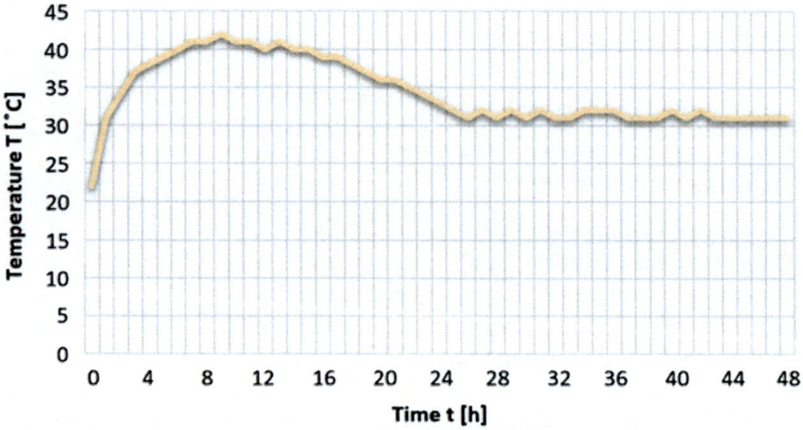

Fig. 5.14 Graphic display of the course of temperature changes during the trial run

and radial direction for the individual BR samples is shown in Table 5.9. Starting torque (Rm) BR TS 050–63, s/n.1911 before the trial run of CW direction is shown in Fig. 5.15 and CWW direction, in Fig. 5.16.

Following the trial run mode and the collecting of the measurements of the reducer's natural frequencies, a summary of the initial information about the bearing reducers TS 050 with a gear ratio of 63, necessary for evaluating the parameter changes, was obtained. Subsequent to the trial run, measurements of the LM, H, and Rm (CW, CCW) parameters were conducted again, as they could have been affected by the trial run mode (Table 5.10). A graphical representation of the course of Rm (CW, CCW) measurement is shown in Figs. 5.17 and 5.18. For the sake of comparability of the individual results before the trial run, after the trial run, and before the actual operation, the measurements of the BR characteristics were conducted in the same position as the support body and the output flange through the mark created in the first measurement. A graphical representation of the course of LM and

Table 5.9 Values of natural BR TS 050 frequencies measured after the trial run (configuration B)

No	Type BR/serial No	Natural frequency radial—torsional [rpm/Hz]
1	TS 050 s/n.1911	820/11.3
2	TS 050 s/n.1912	830/12.0
3	TS 050 s/n.1913	815/13.0
4	TS 050 s/n.1914	835/12.0
5	TS 050 s/n.1915	824/11.5
6	TS 050 s/n.1916	822/13.0
7	TS 050 s/n.1917	838/12.5
8	TS 050 s/n.1918	832/13.0
9	TS 050 s/n.1919	840/11.0
10	TS 050 s/n.1920	837/14.0

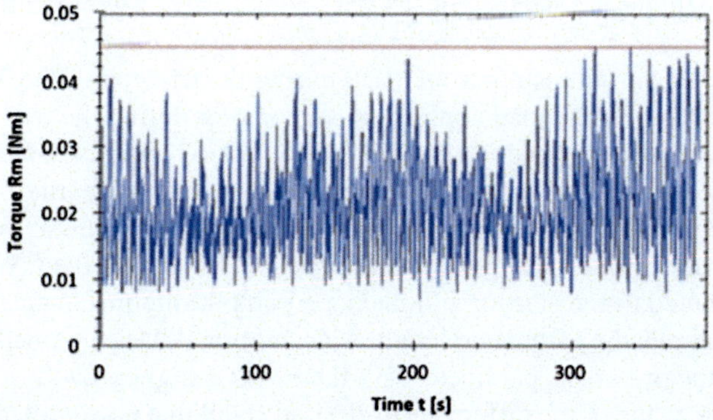

Fig. 5.15 Starting torque (Rm) BR TS 050–63, s/n.1911—CW direction (before the trial run)

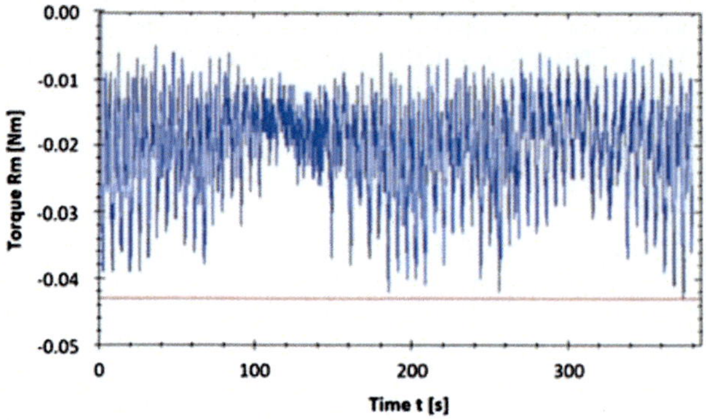

Fig. 5.16 Starting torque (Rm) BR TS 050–63, s/n.1911—CCW direction (before the trial run)

Table 5.10 Bearing reducer TS 050 parameters measured after the trial run (configuration B)

No	Type BR/serial No	Lost motion [arcmin]	Hysteresis [arcmin]	Rm (CW) [Nm]	Rm (CCW) [Nm]
1	TS 050 s/n.1911	0.73	0.59	0.026	0.025
2	TS 050 s/n.1912	0.65	1.00	0.030	0.028
3	TS 050 s/n.1913	0.80	1.13	0.028	0.031
4	TS 050 s/n.1914	0.78	0.65	0.025	0.026
5	TS 050 s/n.1915	0.62	1.05	0.036	0.029
6	TS 050 s/n.1916	0.71	0.84	0.030	0.032
7	TS 050 s/n.1917	0.82	1.08	0.024	0.028
8	TS 050 s/n.1918	0.90	0.75	0.029	0.030
9	TS 050 s/n.1919	0.55	0.70	0.020	0.027
10	TS 050 s/n.1920	0.76	0.93	0.021	0.024

H measurements of a selected BR TS 050–63, s/n.1911 representative is shown in Fig. 5.19.

Summary of partial results from MDS during the trial run of the TS 050 bearing reducer—loading by the load application servomotor during the trial run:

Measurements were conducted on 10 BR TS 050–63 samples. Before the trial run, we performed a series of measurements on the samples to identify the factual state of BR. After an initial 48-h trial run with a load factor of 9 Nm at the output, there was a decrease in the starting torque in both directions, i.e., the CW, CCW, which can be attributed to the starting of kinematic pairs during mechanical transmission. After the trial run, the parameters, such as lost motion (LM), remained unchanged. In the case of the hysteresis parameter (H), there was a slight growth in the increase in value, but the value of the difference itself is so small that no significant effect needs to be attributed thereto. The range of change is 0.01 to 0.05 arcmin (Table 5.11).

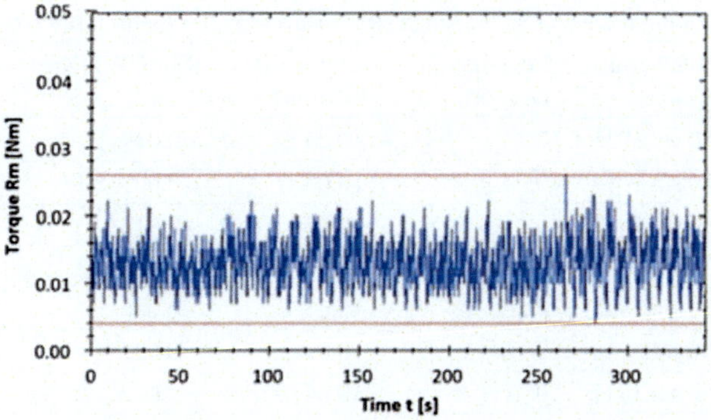

Fig. 5.17 Starting torque (Rm) BR TS 050–63, s/n.1911—CW direction (after the trial run)

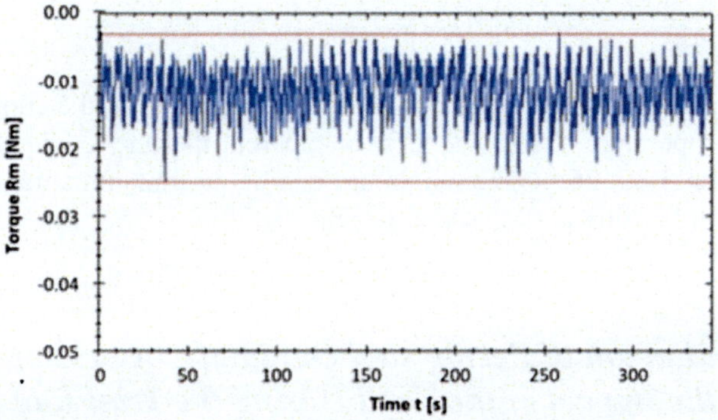

Fig. 5.18 Starting torque (Rm) BR TS 050–63, s/n.1911—CCW direction (after the trial run)

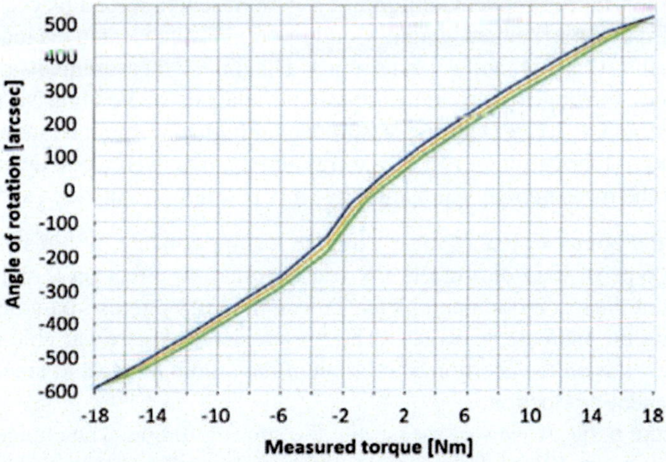

Fig. 5.19 Course of the hysteresis curve BR TS 050–63, s/n.1911

Table 5.11 Bearing reducer TS 050 parameters measured after the trial run (configuration B)

No	Type BR/Serial No	Lost motion [arcmin]	Hysteresis [arcmin]	Rm (CW) [Nm]	Rm (CCW) [Nm]
1	TS 050 s/n.1911	0	+0.04	−0.019	−0.018
2	TS 050 s/n.1912	0	+0.05	−0.005	−0.013
3	TS 050 s/n.1913	0	+0.02	−0.012	−0.015
4	TS 050 s/n.1914	0	+0.02	−0.007	−0.012
5	TS 050 s/n.1915	0	+0.05	−0.006	−0.021
6	TS 050 s/n.1916	0	+0.02	−0.021	−0.014
7	TS 050 s/n.1917	0	+0.04	−0.012	−0.012
8	TS 050 s/n.1918	0	+0.01	−0.015	−0.012
9	TS 050 s/n.1919	0	+0.03	−0.012	−0.012
10	TS 050 s/n.1920	0	+0.03	−0.007	−0.010

From the course of the temperature values measured on the support body after the initial temperature rise to 42 °C, it is apparent that the temperatures stabilized at 31 °C after about 26 h. The maximum degree of contamination of the PG TT1 lubricant with Fe solid particles was 15 ppm Fe.

5.1.5 Evaluation of Results and Qualitative Assessment of the Impact of the Load During the Trial Run Mode

The examined object No. 1 BR TR 050 with a gear ratio of 63 was subjected to two types of the trial run mode. Configuration A represented a load placed on the arm, represented by a mechanical weight. Configuration B represented the load applied by the drive, the so-called motor versus motor. Diagnostic measurements of hysteresis, lost motion, starting torque, temperature and analysis of Fe abrasive particles for Castrol Tribol GR TT1 PD lubricant, suitable for high load applications and designed for long-term lubrication of roller bearings and the like, were done during the trials. Based on the measurements and analyses, it is possible to state the following:

- From the point of view of measuring the lost motion and hysteresis: The choice of the type of load during the BR trial run had no effect on the change in the values of the lost motion parameter. The monitored hysteresis parameters showed a negligible increase in values. The hysteresis curve is a variable nonlinearity and depends on the position of rotation of the output flange and the perfection of component production.
- From the point of view of measuring the starting torque: The choice of the type of load in the BR trial run had a demonstrable influence on the values of the parameter monitored—the starting torque and fluctuation of the starting torque. In the trial run with a load, a sinusoidal load is applied to the kinematic pairs,

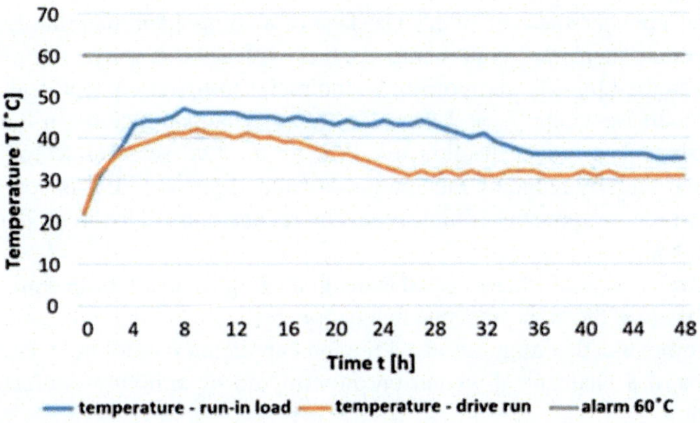

Fig. 5.20 Course of temperatures on the BR TS 050 support body over a 48 h mode

which are evenly loaded by a servomotor with a torque sensor (controlled torque value) during the load application process by the drive. In both trial run methods, the surfaces were smoothed, which was reflected in the torque fluctuation—a drop in the peaks. A decrease in the mean value of the starting torque was recorded when the load was applied by the drive.

- From the point of view of temperature measurement: After the initial increase in the surface temperature BR, the friction regime stabilized, and the temperatures decreased and became stable. The alarm value was not exceeded during the measurement, which can be seen in Fig. 3.18.
- In terms of the lubricant condition: After the initial phase of the relative trial run, the Castrol Tribol GR TT1 PD lubricant used showed a satisfactory Fe content of abrasive particles (all sample values were below the recommended limit) (Fig. 5.20).

5.1.6 Discussion of the Study Mentioned Above

Reduction mechanisms, such as the high-precision bearing reducers of today's drive units in the field of machine tools and automation, are state-of-the-art and technologically advanced. Subjecting such mechanisms to a trial run is important. The development of new materials, lubricants, smaller production tolerances and new final technologies have reduced the trial run length, but it is nevertheless irresponsible to apply the full load to the reduction mechanism right from the start. The obligatory test operation cannot be considered a fully fledged trial run, as the test operation mostly serves the purpose of verifying the functionality of the reduction mechanism.

In the field of mechanical engineering technology of surface production, new technologies enable the production of surfaces to a precisely defined value. Specific properties are characteristic of each surface of a component material. Each bearing reducer component that comes into contact with another surface must adjust its position to the opposite component. In the metal-to-metal contact, friction occurs. Friction increases metal wear. Each reduction mechanism includes rolling elements and bearings that reduce friction by allowing the two surfaces to roll relative to each other, thereby reducing the amount of friction generated. It is now considered a standard for components of high-precision reducers to be deburred, demagnetized and washed.

Research in this area has mapped in detail the impact on all the essential variables' characteristic of the high-precision reducer properties. The trial run was carried out in both recommended configurations of the load application onto the reducers, namely by means of a mechanical weight on the arm and by applying the load through a servomotor.

Based on the research carried out, it is possible to recommend that each user, after installing the bearing reducer in the appropriate alignment, does a trial run during operation for at least 48 h with a 50% TR load under the nominal torque and a maximum speed of 1000 rpm. In case of the oil version, it is recommended to change the entire oil content after the first 100 h of operation following the trial run.

At the same time, based on the research documented above, a methodology was established for measuring the important properties of high-precision reducers in connection with their trial run, which consists of the following steps:

- Reducer trial run over 48 h, max. 60 °C temperature NT, 50% TR at max. 1000 rpm.
- Measurement of abrasion Fe particles—max. Fe content of particles 50 ppm.
- Measurement of the reducer starting torque—input 10 rpm, upon completion of one rotation of the output BR member, at the sampling frequency of the sensor of 200 samples per second.
- Measurement of hysteresis (H) and lost motion (LM)—in load-free state, TR = 100% Nm in both directions (CW, CCW), evaluation at 3% TR load.
- Angular transmission error measurement (ATE)—in load—free state, output 0–720°, in both directions (CW, CCW).
- Repeated accuracy measurement—output rotation 1800°, eight desired positions at 360°.

This methodology unifies our approach to measuring the static and dynamic properties of high-precision reducers to identify their critical parameters and positioning accuracy.

The article points out two possibilities of running-in bearing reducers. The run-in was performed by a mechanical weight on the arm and by a load servomotor. The research showed that the choice of the load has a negligible effect on the values of lost motion and hysteresis, but it has a demonstrable effect on the values of the monitored parameter—starting torque and fluctuation of starting torque. However, in both trial run methods, the surfaces were smoothed, which was reflected in the torque fluctuation—a drop in the peaks. At the same time, after extensive research, it can be stated that the correct running-in of bearing reducers after their commissioning means the improvement of all parameters that characterize the static and dynamic parameters of high-precision reducers.

5.2 Design and Implementation of a Diagnostic System for Measuring High-Precision Reducers

The current industry requirements for driving units place increased demands on the transmission of force effects with higher kinematic accuracy, as well as on the precise setting of the position of these effects. This results in deployment of high-precision cycloidal and harmonic propulsion systems.

The main part of these systems is a reducer the main function of which is to transform power quantities. It is designed to reduce the speed of the driving machine to the input speed of the driven machine at constant transmission.

These high-precision transmission mechanisms allow the use of reducers in robotics and automation, machine tools, measuring devices, navigation systems, in the aerospace industry, in the military and medical field, in woodworking, in the printing industry, in textile and glass processing machinery, in filling machines, etc (Fig. 5.21).

Technological complexity is the hallmark of reducers production with high demands on accuracy and constitutes a niche engineering production, as its outputs must meet the strictest international criteria for production accuracy, assembly, measurement of selected parameters (accuracy characteristics), durability and reliability tests. The production of individual components consists of roughing (blasting) operations, heat treatment, final (grinding) operations and the assembly itself, and last but not least, measurement of functional dimensions and kinematic quantities. From the point of view of final (finishing) operations, rounding the external and internal dimensions necessitates the application of the most recent grinding technologies. The manufacturing accuracy of the internal and external dimensions of the components is measured in microns. In most cases, single-purpose devices are applied, considering

Fig. 5.21 Axial decomposition of the TwinSpin reducer by Spinea s.r.o

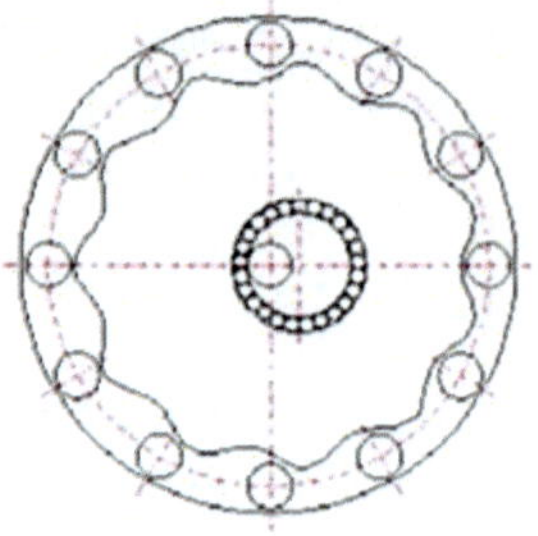

Fig. 5.22 Cycloidal transmission

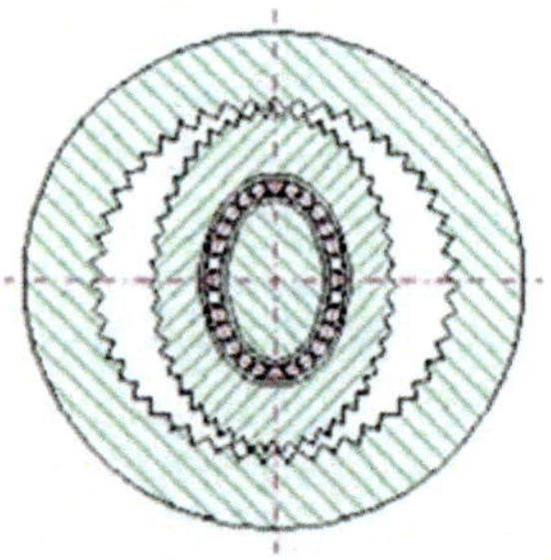

Fig. 5.23 Harmonic transmission

the design and technological requirements placed on the component, as well as on the superiority of assembly.

High-precision reducers are based on several mechanical transmission principles:

- Planetary mechanism—reducers with involute gearing (Nabtesco).
- Cycloid mechanism—reducers with trochoidal gearing (Sumitomo, Spinea).
- Harmonic mechanism—reducers with involute gearing (Harmonic Drive) (Figs. 5.22 and 5.23).

Despite many positive features, the reducer also exhibits its characteristic nonlinearities (e.g., nonlinear mechanical characteristics such as lost motion, angular transmission error, interference with higher harmonics, influence of interpolation distortion), which need to be further investigated and analyzed, as they often have a significant impact on the behavior of the system as a whole. Compensation for these nonlinearities subsequently beckons the possible use of any linear regulator. The use of a suitable mechatronic diagnostic system is necessary for correct identification of and subsequent compensation for these nonlinearities [28].

5.2.1 *Design of a Mechatronic Diagnostic System for Measuring High-Precision Reducers*

To create a general idea of the process of analysis of the mechatronic diagnostic system of the object (high-precision reducers) examined, it is necessary to create the model defined by the object, as well as models of diagnostic mechanisms. A mechatronic diagnostic system can generally be characterized as a mechatronic system interconnected through functional elements that respond independently to a set of input signals. The functional elements and the links between these elements form a set represented by a diagram of the diagnosed object.

The aim of the solution is to achieve a functionally and structurally integrable diagnostic process of the mechatronic system. Such representation of the system consists of a set of elements and steps corresponding to the research and development purpose and diagnostic tasks of the object under examination (Fig. 5.24).

We will use the model to study the relational structure and behavior of the mechatronic system via a controlled system of monitoring and diagnosing the investigated object. The methods, procedures, and means of choice focus on monitoring and troubleshooting the technical condition of the examined object. The result is a diagnosis that makes it possible to make recommendations for production, operation, and maintenance. Such clarification of the nature of elements affecting reliability can be

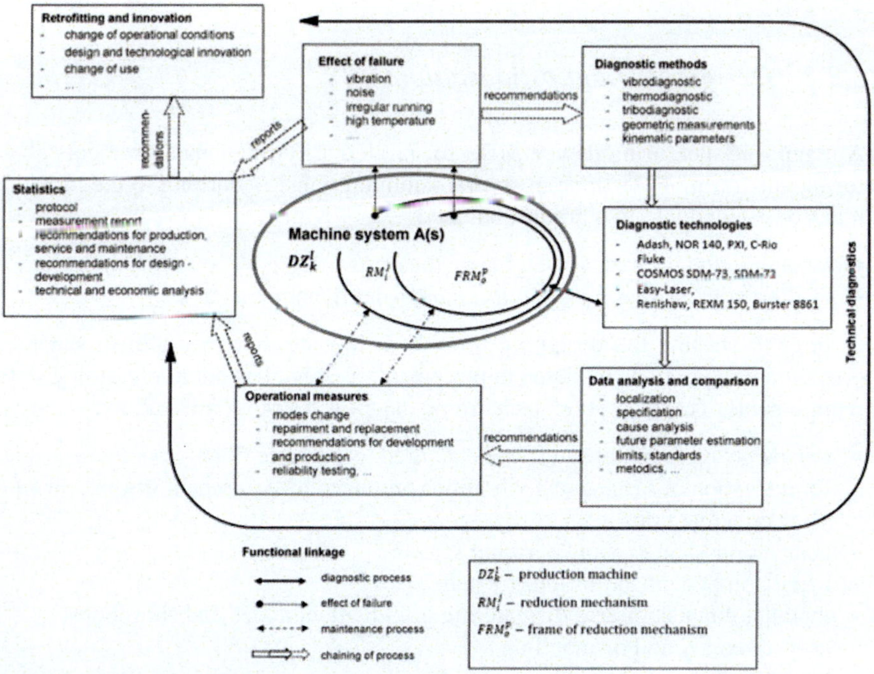

Fig. 5.24 Mechatronic diagnostic system design

further used in changing operating conditions, as well as in design and technological innovation.

Parameters of a specific technical design and implementation of a modular diagnostic device, in which the system of measuring devices and components necessary for analysis and evaluation will be integrated, need to be defined as follows:

- diagnosed object—type, size,
- monitored quantity—parameter, range (measured technical parameters).

The investigated object to be diagnosed is a reducer or a rotary actuator and the rolling bearings. Harmonic Drive, Nabtesco, Sumitomo, and Spinea are among the world's leading manufacturers of high-precision reducers used in robotics and equipment requiring high-precision rotary motion. The products made by the respective manufacturers differ both in principle and in the company itself [29].

The mechatronic diagnostic system proposed will make it possible to conduct the following measurements and tests:

- measuring the LR effectiveness,
- measuring the LR starting moments,
- measuring the LR feedback moments,
- measuring LR vibrations and temperature at defined work cycles,
- start-up tests,
- LR durability tests.

5.2.2 Design of Diagnostic Equipment

A diagnostic measuring device MDS for particular types of measurements allows for the measuring to be done in several configurations. According to the load type, it can be divided into two groups, namely:

- loading with mechanical weights on the arm,
- loading with a stress servomotor or a powder dynamometer.

Prior to starting the designing process, it is necessary to establish secondary operational requirements related to placement, use and, last but not least, legislative requirements. Therefore, the mechatronic diagnostic system must allow:

- to have its base part of the frame anchored to concrete floor,
- for insulation of mechanical vibrations produced by equipment positioned on the same concrete floor,
- connection to an existing earthing system,
- metal components be surface treated,
- modular connection and dismantling of individual nodes and subgroups,
- easy assembly and dismantling,
- the space for rotational movement of loads be secured with a mechanical barrier,

- the area be secured against entry of unauthorized persons (e.g., with a mechanical barrier—this is not a part of the design).

The mechatronic diagnostic system will consist of three basic parts:

- Mechanical structure—frame.
- Control part.
- Measuring part.

The measuring device design consists of a set of interconnected mechanical components. The structure consists of a free-standing frame, a linear system and fastening flanges and shafts.

The final design of the mechatronic diagnostic system for measuring the bearing reducer (Fig. 5.25) consists of a support structure made of a frame (1), a linear guide (2), a bracket (3), (4) and (5), cranks (28) and (29). The structure consists of four nodes: the stress node, the drive node, the sensing node, and the test node. The stress node consists of a flange (6), an intermediate flange (7) and a stress drive (8) and provides for the creation of a load moment. The drive node generates movement according to the parameters required for making the measurements and consists of a flange (22), a drive (23), a separate linear guide (24), and a crank (25).

The sensing node ensures the measurement of torque and angular velocity values and consists of an output torque sensor (10), an input torque sensor (20), a stand (26) and (27). The test node consists of an output flange (14), a bearing reducer (15), an intermediate flange (16), a flange (17), a shaft (18). Linking of individual nodes into the measuring chain is done by means of a coupling (9), a bellows coupling (11), a reducer (12), a quick coupling (13), a bellows coupling (19), and a coupling (21). Measurement control and sensing of the measured quantities is ensured by the input / output device (33).

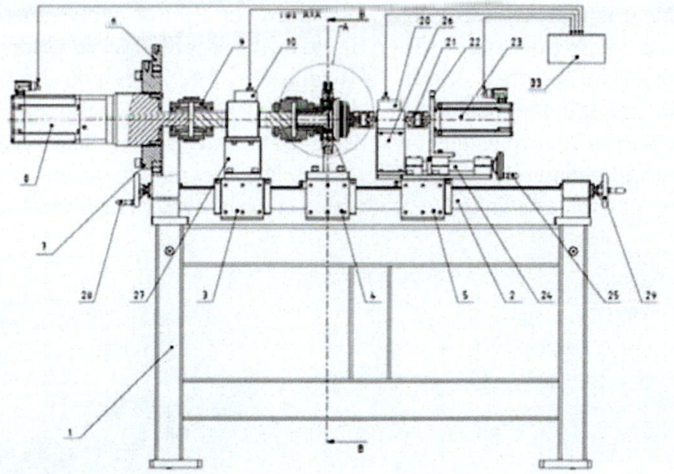

Fig. 5.25 Diagnostic system—structural design

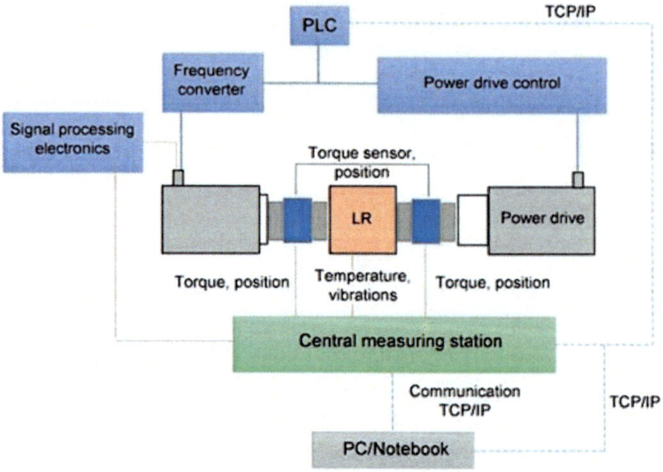

Fig. 5.26 Principle connection diagram of diagnostic system

An integral part of this device are backlash-free bellows couplings, which are used, among other things, for their ability of backlash-free synchronous transmission of rotary motion and the ability to compensate for manufacturing inaccuracies.

Hydraulic ETP Express type clamping sleeves are used for a reliable connection of the shaft to the hub. In addition to the transmitted torque, crucial criteria include properties such as reliability of connection, fast clamping, and connection release.

The structural design was followed by design of the mechatronic diagnostic system's control and measuring part. The proposed circuit diagram of the diagnostic system for measuring bearing reducers is shown in Fig. 5.26.

In industrial applications, as well as in our case where we need to regulate speed, the electrical equipment most frequently used for control and regulation of drive mechanisms is a frequency converter. Based on the specified servomotor power series derived from the specified objects of diagnostics, frequency converters to be used for measurement in the mechatronic diagnostic system were designed (Fig. 5.27).

A type 8661 torque transducer was designed to measure static and dynamic torques clockwise and counterclockwise (Fig. 5.28). It is designed for contactless operation and zero maintenance adapted to production environments. The torque sensor 8661 is

Fig. 5.27 Proposed frequency converter

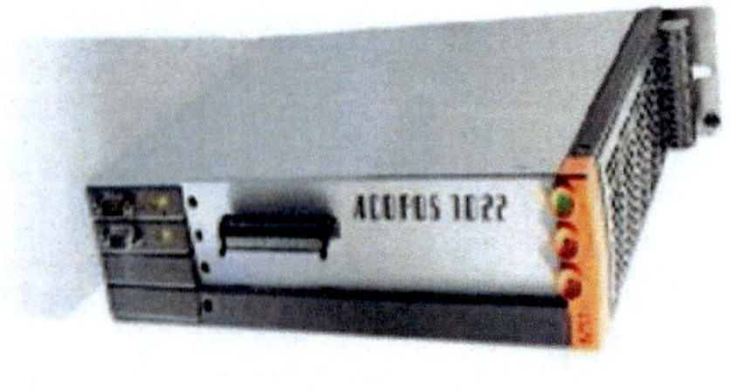

Fig. 5.28 Torque sensor 8861

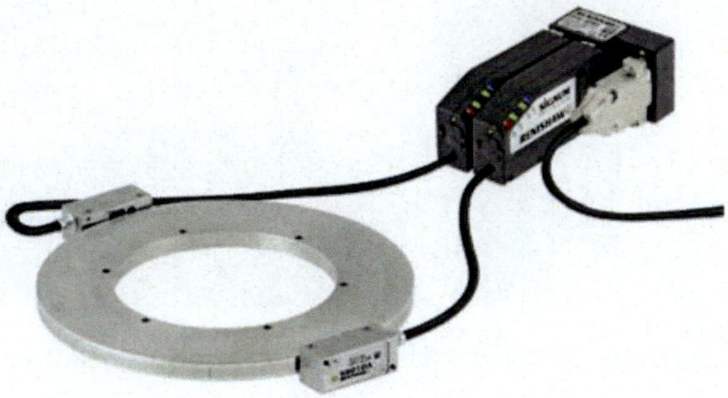

Fig. 5.29 Renishaw position sensor, REXM 150

a special-purpose device for measuring constant torque during continuous operation and measuring torque, tear-off torque. Built-in USB interface and connection to a laptop with the DigiVision PC program makes instant digital access to torque measurement diagnostic data possible. DigiVision PC allows us to read, visualize, and archive the data measured. The 8661-torque sensor enables the system integration into the LabView environment.

For high accuracy of position measurement at the output of the bearing reducer in both directions, a position sensor of REXM 150 type from Renishaw will be used (Fig. 5.29).

The sensor has a pair of reading heads to eliminate errors or deviations caused by inaccuracy in the mounting of the measuring disc. The sensor emits an output analog signal ± 10 V and a pair of incremental TTL signals. Suitable for applications that require the highest angular accuracy, the REXM angular position sensor is ideal for metrological purposes of measuring angular accuracy.

5.2.3 *Results and Discussion*

The present study deals with the design of a diagnostic system for measuring high-precision reducers and rotary actuators. The rotary actuator is a higher unit that integrates the reduction mechanism, the servomotor, and the position sensor. Prior to the design itself, the main and the specific goals were set, the aim of which was

Fig. 5.30 A real view of the constructed mechatronic diagnostic system

to design a mechatronic diagnostic system for research and development purposes and diagnostic tasks. One of the requirements was to design a modular diagnostic device capable of adapting to defined troubleshooting tasks, changing the measured quantity or the size of the transmission mechanism. At the same time, requirements have been voiced for the system to be designed in a way that, if necessary, it could easily change the measured configuration and, at the same time, allow for integration of other measuring devices and instruments.

Subsequently, an MDS diagnostic system model and configuration of measurements according to the type of load and the quantity measured were designed. Prior to the diagnostic system's actual implementation, a draft design was made in Creo. The design considered dimensional and power parameters of the investigated objects serving as the basis for deriving the conversion of power series of servomotors, converters, sensors, as well as torque transmission components, connection brackets and shafts. Subsequently, a diagram connecting specific hardware with software was elaborated. Data flow control was processed in the programming environment of Labview.

The resulting mechatronic diagnostic system is fully functional, and it is used for measuring and diagnosing high-precision bearing reducers (Fig. 5.30).

References

1. Baron, P.—Kočiško, M.—Dobránsky, J.—Pollák, M.—Cmorej, T.: Research and Correlation of Diagnostic Methods for Assessment of the State of Oil Filling in Cycloid Gearbox—2015.In: Advances in Materials Science and Engineering. P. 97841–97841.—ISSN 1687-.
2. Baron, P.—Kočiško, M.—Dobránsky, J.—Pollák, M.—Töröková, M.: Research and application of methods of technical diagnostics for the verification of the design node—2016.In: Measurement. Vol. 94 (2016), p. 245–253.—ISSN 0263-2241.
3. Kočiško, M.—Pollák, M.—Töröková, M.—Baron, P et al.: Determination of Methodology and Research of the Influence of the Trial Run of High-Precision Reducers on the Change of Their Characterizing Properties—2021.In: Applied sciences.—Bazilej (Švajčiarsko) : Multidisciplinary Digital Publishing Institute vol. 11, no. 9 (2021), pp. [1–23].—ISSN 2076–3417.
4. Sorin-Dumitru, G.—Vasile, C.: Industrial automation in the relative spiral feed and contact contour during cold orbital deformation. Int. Conf. Qual. Innov. Eng. Manag. 2012, 13, pp. 367–370.
5. Krolczyk, G.—Legutko, S.—Stoic, A.: Influence of the cutting parameters and conditions onto surface hardness of duplex stainless steel after turning. Teh. Vjesn. Tech. Gaz. 2013, 20, 1077–1080.
6. Pham, A.D.—Ahn, H.J.: High precision reducers for industrial robots driving 4th industrial revolution: State of arts, analysis, design, performance evaluation and perspective. Int. J. Precis. Eng. Manuf.-Green Technol. 2018, 5, 519–533, doi:https://doi.org/10.1007/s40684-018-0058-x.
7. Blagojevic, M.—Marjanovic, N.—Djordjevic, Z.—Stojanovic, B.—Disic, A.: A new design of a two-stage cycloidal speed reducer. J. Mech. Des. 2011, 133, 085001, doi:https://doi.org/10.1115/1.4004540.
8. Sun, X.—Han, L.—Wang, J.: Tooth modification and loaded tooth contact analysis of China bearing reducer. Proc. Inst. Mech. Eng. Part C J. Mech. Eng. Sci. 2019, 233, pp. 6240–6261, doi:https://doi.org/10.1177/0954406219858184.
9. Mao, H.—Liu, G.—Zeng, D.—Cao, Y.;—He, K.—Du, R.: A new design of cycloidal planetary reducer with internal cycloidal profile. ASME 2018 Int. Mech. Eng. Congr. Expo. 2018, pp. 1–8, doi:https://doi.org/10.1115/IMECE2018-86184.
10. Huang, Y.—Du, X.—Zhu, C.—Ni, G.—Ullah, N.—Liu, H.: Mesh stiffness analysis of beveloid gears for the rotating vector transmission. J. Mech. Sci. Technol. 2019, 33, pp. 3943–3953, doi:https://doi.org/10.1007/s12206-019-0739-6.
11. Pham, A.D.—Ahn, H.J.: Efficiency analysis of a cycloid reducer considering tolerance. J. Frict. Wear 2017, 38, 490–496, doi:https://doi.org/10.3103/S1068366617060113.
12. Terakawa, T.—Komori, M.—Morita, Y.: Motion analysis and experiments of reducer-integrated motor using strain wave gearing system. In Proceedings of the ASME 2018 International Design Engineering Technical Conferences and Computers and Information in Engineering Conference, Quebec, Canada, 26–29 August, 2018; pp. 1–6, doi:https://doi.org/10.1115/DETC2018-85299.
13. Gorla, C.—Davoli, P.—Rosa, F.—Longoni, C.—Chiozzi, F.—Samarani, A.: Theoretical and experimental analysis of a cycloidal speed reducer. J. Mech. Des. 2008, 130, 112604, doi:https://doi.org/10.1115/1.2978342.
14. Yamada, S.—Fujimoto, H.: Precise joint torque control method for two-inertia system with backlash using load-side encoder. IEEJ J. Ind. Appl. 2019, 8, pp. 75–83, doi:https://doi.org/10.1541/ieejjia.8.75.
15. Shih, Y.P.—Lo, S.H.—Sheen, B.T.—Tsai, Y.H.: Precision evaluation for cycloidal gears. In Proceedings of the ASME 2019 International Design Engineering Technical Conferences and Computers and Information in Engineering Conference, California, USA, 18–21 August, 2019; pp. 1–5, doi:https://doi.org/10.1115/DETC2019-97327.

16. Executive Summary World Robotics 2016 Industrial Robots. Available online: http://www.ifr.org/fileadmin/user_upload/downloads/World_Robotics/2016/Executive_Summary_WR_Industrial_Robots_2016.pdf (10. April 2020).
17. Xiong, Y.—Huang, K.—Xu, F.—Yi, Y.—Sang, M.—Zhai, H.: Research on the influence of backlash on mesh stiffness and the nonlinear dynamics of spur gears. Appl. Sci. 2019, 9, 1029, doi:https://doi.org/10.3390/app9051029.
18. Margielewicz, J.—Gąska, D.—Litak, G.: Modelling of the gear backlash. Nonlinear Dyn. 2019, 97, 355–368, doi:https://doi.org/10.1007/s11071-019-04973-z.
19. Getachew, A.A.—Hirpa, G.L.: Effect of backlash on transmission error and time varying mesh stiffness. IWAMA 2020 Adv. Manuf. Autom. X 2021, 737, 18–28, doi:https://doi.org/10.1007/978-981-33-6318-2_3.
20. Spitas, C.—Spitas, V.: Coupled multi-DOF dynamic contact analysis model for the simulation of intermittent gear tooth contacts, impacts and rattling considering backlash and variable torque. J. Mech. Eng. Sci. 2016, 230, 1022–1047, doi:https://doi.org/10.1177/0954406215596696.
21. Walha, L.—Fakhfakh, T.—Haddar, M.: Nonlinear dynamics of a twostage gear system with mesh stiffness fluctuation, bearing flexibility and backlash. Mech. Mach. Theory 2009, 44, 1058–1069, doi:https://doi.org/10.1016/j.mechmachtheory.2008.05.008.
22. Xue, J.—Qiu, Z.—Fang, L.—Lu, Y.—Hu, W.: Angular measurement of high precision reducer for industrial robot. IEEE Trans. Instrum. Meas. 2021, 70, 1–6, doi:https://doi.org/10.1109/TIM.2020.3028887.
23. Caldarola, F.—Maiolo, M.—Solferino, V.: A new approach to the Z-transform through infinite computation. Commun. Nonlinear Sci. Numer. Simul. 2020, 82, 105019, doi:https://doi.org/10.1016/j.cnsns.2019.105019.
24. Farshidianfar, A.—Saghafi, A.: Global bifurcation and chaos analysis in nonlinear vibration of spur gear systems. Nonlinear Dyn. 2014, 75, 783–806, doi:https://doi.org/10.1007/s11071-013-1104-4.
25. Truxal, J.G. Control Engineers' Handbook: Servo-Mechanisms; McGraw-Hill: New York, NY, USA, 1958; p. 1112.
26. Xu, H.—Shi, Z.Y.—Yu, B.—Wang, H.: Optimal measurement speed and its determination method in the transmission precision evaluation of precision reducers. Appl. Sci. 2019, 9, 2146, doi:https://doi.org/10.3390/app9102146.
27. Hric, M.—Durovsky, F.—Fedak, V.: Influence of non-linearities of cycloid gearbox. ATP J. 2012, 9, 42–44.
28. Gorla, C.—Davoli, P.—Rosa, F.—Longoni, C.—Chiozzi, F.—Samarani, A.: Theoretical and experimental analysis of a cycloidal speed reducer. Journal of Mechanical Design, 2008, 130, 1–8, doi: https://doi.org/10.1115/1.2978342.
29. Executive summary world robotics 2016 industrial robots. Available online: http://www.ifr.org/fileadmin/user_upload/downloads/World_Robotics/2016/Executive_Summary_WR_Industrial_Robots_2016.pdf.

Chapter 6
Conclusion

The role of processes arising from implementation of technical diagnostics is based on objective knowledge of the technical condition of monitored objects and particularly in ensuring the capabilities of the technical system to perform required functions under stated conditions. In the actual sense, the technical diagnostics represents discipline that follows the forms of manifestating of failures and developing of methods for detecting them. [1, 2]

The most important role of technical diagnostics consists in predicting the technical state of the system under review. One aim of the technical diagnostics is to predict the duration of trouble-free operation that means the time interval from the reference point to the time of failure. The role of forecasting failures rests in identifying the deterioration of system under review and determination of the moment when it is necessary to implement appropriate measures in order to prevent failures or damages. [1, 2]

Where the condition of the system under review is being marked with parameter S, it is represented by the time-dependent function of S (t). Through the known values of S (ti) for discrete time points ti(i = 0,1,2,3,…,k) at the time of commissioning the system until the moment of diagnosis tk, it is possible to predict function values S(tk + 1), S(tk + 2), …, S(tk + j), …, S(tk + m) for time moments tk + 1, tk + 2, …, tk + τ (Fig. 6.2). [1, 2] (Fig. 6.1)

The period of achieving the critical value can be determined by assessing the last two conditions, i.e., condition during diagnosing S(tk) and condition during previous diagnosing S(tk-1). The failure occurs at the moment tk + τ, i. if S(tk + τ) = Sm. [3]

Determination of the time of critical value Sm can be determined from the relation [1, 2]:

$$t_k + \tau = \frac{S_m - S(t_k)}{S_m} = \frac{(S_m - S(t_k)).(t_k - t_{k-1})}{S(t_k) - S(t_{k=1})} \tag{6.1}$$

P. Baron et al., *Application of Troubleshooting Tools in the Monitored Production Processes*, Management and Industrial Engineering,
https://doi.org/10.1007/978-3-031-41428-2_6

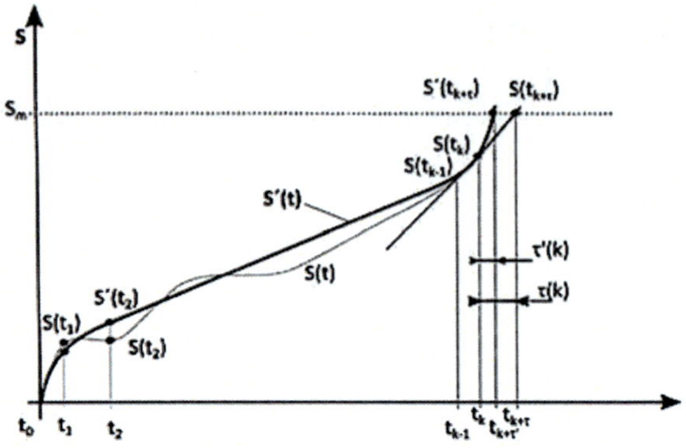

Fig. 6.1 The course of dependence of diagnosed parameter on time [3]

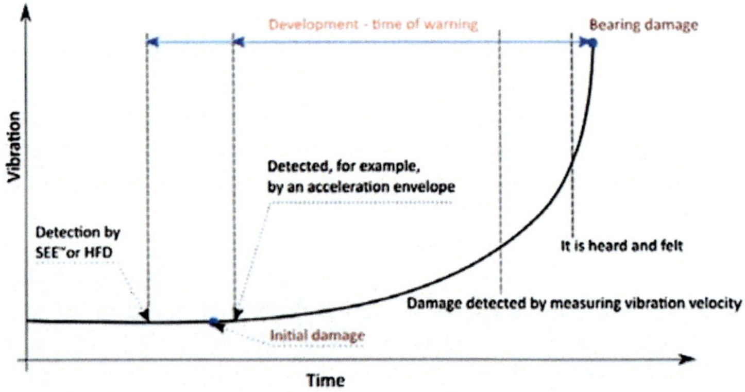

Fig. 6.2 Bearing damage evolution—the trend of measured rolling bearings vibrations [5, 6]

Such a methodology is applicable in case of the existence of short diagnostic intervals. In the case of longer intervals that relationship may not represent the nature of wear, especially for reasons [1, 2]:

- The rate of wear at the end of the technical life-cycle increases, i.e., failures may occur sooner, as it is clear from the above relationship
- With identical values of parameters S(tk) and S(tk-1) there is a division by zero—the equation cannot be used (this may be the case if the metering device is not satisfactory, mainly because of the practical reading of values measured).

When determining the residual operational capability there is required for the above reasons to consider not only the last two values of real function S(t), but the whole course of wear—all parameters from the interval < S(t1), S(t2), …, S(tk-1), S(tk) > at the moment tk and by correlation analysis to determine the functions S'(t). In this case, the failure will be predicted at the time tk + τ', t.j. ak S'(tk + τ') = Sm [1, 2].

Even at present, using modern tools of maintenance, the arising failure occurs often and very shortly before the outbreak of failure when maintenance is too busy to carry out repairs, because companies supplying spare parts, respectively providing repairs, often do not take into account the specific conditions of concrete operation.

To increase the reliability of the device is therefore necessary to carry out the following requirements [1–3]:

- Increasing the time between repairs (the ability to be ready for operation),
- Shorten repair time,
- Early repairs to reduce the consequences of failure.

To meet these requirements, there must be available information on what is the real condition of the device. This way can be removed also hidden, less serious errors, which may later develop into larger, often catastrophic failure. On the basis of this information, we may intervene against failure in the most appropriate time and in terms of errors (error range and its process) and the impact on the production process (e.g.: during weekends, holidays, when production does not take place on the device). By the mentioned process, there can be removed not only the consequences but also early signs of failures and this way it is possible to significantly increase the lifetime of the technical system under review [2].

The fundamental requirement of modern maintenance is early diagnosis of failures. Even today, there is a frequent occurrence of the fact that especially small businesses do not have the technical diagnostics established at all and the only diagnostic method is represented by subjective evaluation of machines, respectively by analysis of machine after failure. In the field of analysis of lubrication and lubricants, there are no tribotechnicians in such companies at all, so the lubricating of machines is thus represented by a random process. In large companies, the situation is different. Technical diagnostic is present in some form, but it happens often that the results are not well enough interpreted. This is often the issue especially of the human factor connected with a lack of data integration from different sources into maintenance management system [2].

Lubricants can bring considerable impact on the operation of machines and equipment, on their reliability which often depends on the proper selection and care of lubricants during their operation. From the literature data that are confirmed by the practice is known that up to 49% of gearbox damages are due to reason of bearing damage, 41% by reason of damage to the teeth and 10% are given as other reasons. The reasons must be found in the fact that it is a high-pressure systems (Hertz pressure) resulting in the formation of small fraction particles that cause bearing damage. It should be mentioned the new phenomenon of wear on the gear (micro pitting) generated on the tooth surface and causes a change in the profile of the tooth and leads to a higher noise level of gearing and causes the other fatigue wear (macro pitting, flaking, etc.). Micro pitting may also occur in bearings. The effect of the metal particles of oil degradation depends on the type of particles (Fe, Cu and others). We know that these particles may evoke a catalytic oxidation. One of the requirements is the regular inspection of the particles. It is a process which results from the monitoring of the friction parts, machine nodes and from the fact that the machine makes its own particles itself and therefore the requirement for periodic care of lubrication, in order to maintain a balance of particles present in the lubricating oil.

Rolling bearings are important parts of machines, mechanisms, and devices that enable them to operate reliably and provide for the safe and long-term functioning of the technical systems without friction and enormous wear. On the other hand, roller bearing damage is an important and significant phenomenon that can lead to the failure of machines, mechanisms, and equipment in all industrial areas where they are used. The analyses of the breakdown causes, their possible consequences, and their removal create conditions for ensuring the quality and reliability of rolling bearings during their technical life. Similar to other machine components, even in the case of roller bearings, there may be a variety of causes of premature damage and failure of the bearing. It is necessary to differentiate between bearing durability, determined by wear due to the application of load at operating speed, and bearing lifetime, which is a period of the bearing serviceability before it is for various reasons decommissioned. The life of the bearing is affected, for example, by improper installation, misalignment of bearings, by manufacturing defaults in the production of connecting parts, by improper handling of bearings, by impurities deposited in bearings, or by poor lubrication. When damage or other deficiencies occur in bearings, it is important to determine the cause so that measures can be taken to prevent further damage [4, 5].

The research into bearing damage includes the knowledge of metallurgy, tribology, operating environment, and consideration of the load applied. In some cases, even a single event may cause malfunctioning. Sometimes it requires a combination of several complex events and circumstances, in addition supported by a faulty human decision, adverse environmental conditions, demanding operating conditions, and improper maintenance to make such an event occur. All of this contributes to premature failure of bearings. Ultimately, this leads to a disastrous breakdown of the rotating component or of a component moving in a reciprocal manner, and to additional damage to the associated mechanisms. Due to the high cost of production losses and due to the time constraints that the service department experiences, the reality check calls for a maximally simplified and effective problem-solving process [4, 5].

The main problem of diagnostics is the determination of critical locations in which a suitable electrical output sensor can measure the change in the physical variable characteristic of the fault, its origin, and development. The measured change of one or more physical variables can serve as a means of determining a reliable criterion that would characterize the origin and development of the fault up to the moment of the breakdown, as well as the time period during which the object can be used and the time point when it has to be dismantled and repaired (Fig. 6.2). In practice, it is not always possible to monitor all the operating parameters of the device with complex and expensive modern measuring devices. Therefore, even in today's technically advanced times, in some cases, it is advisable and useful to monitor the noise, temperature rise, or vibration, which may indicate a bearing failure. Bearings that are in good condition produce fine humming when in operation. Screeching or grinding sounds indicate that the bearing is not in a good condition. If the bearing operates for a longer time at about 125 °C, its durability is reduced. It is visually checked whether the bearing is sufficiently lubricated, well-sealed, or not visibly worn [3, 5]. Several methods and techniques successfully have been analyzed where the direct influence on performance and tribological properties of the product have been defined [5].

As one of the most common mechanical equipment classes, rotation machinery occupies an important role in industrial applications such as manufacturing, metallurgy, energy, and transportation. Due to tough working environments, similar materials, and structural properties, rotation machinery can be subject to malfunctions or failures. This can significantly decrease machinery service performance including manufacturing quality and operation safety and cause machinery to break down, which may lead to serious catastrophes. Accordingly, research into fault diagnosis of rotation machinery has attracted considerable attention by researchers in related domains in recent years. The vibration signals collected from velocity or accelerator sensors located in machinery housing are generally regarded as the foundation of fault diagnostic procedures. However, most existing studies on fault diagnosis of rotation machinery have empirically or experimentally focused on analyzing single sensor signals [5].

In industrials, vibration-based diagnosis using accelerometers mounted on the gear/bearing housing has prevailed in the recent decades because it does not interfere with the normal operation of rotating machines [5].

In terms of signal processing, vibration signals are usually employed to analyze the rotating machinery faults because the vibration signals are easy to acquire and highly correlate with the condition of the rotating machinery. However, the vibration signals contain high-dimensional data and are enclosed by a lot of irrelevant and redundant information, which degenerate the accuracy as well as the fault identification time of the diagnostic system. To solve the problem, a proper feature extraction technique which can extract useful information of faults from the vibration signal is desirable. At present, there exist many methods to extract the features of faults from vibration signals, such as Fourier transform, short time Fourier transform, and wavelet transform [5].

References

1. Balog, J. - Chovanec, A. – Kianicová, M.: Technical Diagnostics, TnUAD, Trenčín, 2003
2. Baron, P. - Kočiško, M. - Dobránsky, J. - Pollák, M. - Töröková, M.: Research and application of methods of technical diagnostics for the verification of the design node - 2016.In: Measurement. Vol. 94 (2016), p. 245–253. - ISSN 0263–2241
3. Panda, A. - Prislupčák, M.: Valuation of vibrations during machining, Trends and Innovative Approaches in Business Processes, 17. International Scientific Conference, Almanac, Košice, 2014, pp. 1–5
4. Salokyová, Š. - Ružbarský, J. - Krenický, T.: Vibration diagnostics in manufacturing system with water jet technology, RAM-Verlag Lüdenscheid, Germany, 2015, p. 118

5. Baron, P. - Kočiško, M. - Dobránsky, J.: The dynamic parameters correlation assessment of the textile machine high-speed bearings in changed technological conditions - 2018.In: Measurement. Vol. 116 (2018), p. 575–585. - ISSN 0263–2241
6. Beneš, Š. – Tomeh, E.: Methods of diagnosis of rolling bearings, VŠST Liberec (1991), p. 55

Index

P. Baron et al., *Application of Troubleshooting Tools in the Monitored Production Processes*, Management and Industrial Engineering,
https://doi.org/10.1007/978-3-031-41428-2